Global Warming: Truth and Consequences

Frank S. Levin

Global Warming: Truth and Consequences

A Primer

 Springer

Frank S. Levin
Department of Physics
Brown University
Providence, RI, USA

ISBN 978-3-031-27022-2   ISBN 978-3-031-27023-9   (eBook)
https://doi.org/10.1007/978-3-031-27023-9

© The Editor(s) (if applicable) and The Author(s), under exclusive license to
Springer Nature Switzerland AG 2023
This work is subject to copyright. All rights are solely and exclusively licensed by
the Publisher, whether the whole or part of the material is concerned, specifically
the rights of translation, reprinting, reuse of illustrations, recitation, broadcasting,
reproduction on microfilms or in any other physical way, and transmission or
information storage and retrieval, electronic adaptation, computer software, or by
similar or dissimilar methodology now known or hereafter developed.
The use of general descriptive names, registered names, trademarks, service marks,
etc. in this publication does not imply, even in the absence of a specific statement,
that such names are exempt from the relevant protective laws and regulations and
therefore free for general use.
The publisher, the authors, and the editors are safe to assume that the advice
and information in this book are believed to be true and accurate at the date of
publication. Neither the publisher nor the authors or the editors give a warranty,
expressed or implied, with respect to the material contained herein or for any errors
or omissions that may have been made. The publisher remains neutral with regard
to jurisdictional claims in published maps and institutional affiliations.

This Springer imprint is published by the registered company Springer Nature
Switzerland AG
The registered company address is: Gewerbestrasse 11, 6330 Cham, Switzerland

ACKNOWLEDGEMENTS

Global Warming: Truth and Consequences is an outgrowth of the many courses on global warming and climate change I have taught in various adult education venues. The course content has evolved over time, with the current version being the basis for this book. I am grateful to the comments and questions raised by the students in these classes: they have helped to vivify the lectures, and I am pleased to acknowledge their input.

The preface, the introduction, and the chapters were read and critiqued by Leland Jackson and Carol Levin. I am deeply indebted to them, since their efforts not only led to the eradication of errors and improvements in clarity, but also made me focus even more on how to guide readers without a science background through the complexity of some of the topics.

It is a pleasure to thank Robert Doe for his decision to publish this book and for his clarifying comments concerning publication, and Arumugam Deivasigamani for his elucidating comments before and during the production of it.

Helpful comments on some of the material by Jay O'Neil and Jill Rinkel are gratefully acknowledged.

I am appreciative to Makiko Sato for the explanation on how to understand the new version of the global temperature anomaly graphs on the web site Updating the Climate Science.

Special thanks goes to the persons who gave permission to use the various figures and the information in the tables that are listed in the bibliography: James Hansen, Barton Paul Levenson, Robert Monroe, and Makiko Sato.

As with my other books, it is again my great pleasure to dedicate *Global Warming: Truth and Consequences* to my beloved wife Carol Levin. She not only read and critiqued the manuscript, but she also provided the support and atmosphere needed to complete it. I am grateful to her in countless ways.

PREFACE

Among the many significant changes that have occurred over the last hundred years are the very large increase of average life expectancy globally and the technological developments that not only contributed to it but also altered the ways people live, work, travel, communicate, and respond to the world around them, especially in developed countries.

This increase in life expectancy is mainly a result of greatly reduced mortality rates among babies and young children, which has also led to enormous growth in the world's human population, from approximately 1.9 billion people in 1920 to roughly 7.7 billion in 2021. That such growth could occur is due to a vast increase in food production coupled with an abundant supply of water for agriculture and other human uses. Unfortunately, as discussed in this book, there is a well-founded likelihood that the current food and water availability, which is already compromised, will become even more so in the coming decades.

To understand the origin of this likelihood one must go back to the coal-based industrial revolution of the 1700s that began in England and soon spread to other parts of the world. Prior to this time, the use of coal was largely limited to the cooking of food and the heating of dwellings in colder climates, although wood was used for these purposes as well. But during the industrial revolution coal burning dramatically increased since it was used to power the steam engines that had been invented then. In modern times, coal burning is mainly used to generate electricity, supplemented for this purpose by natural gas and oil, fuels that are outcomes of the technological advances noted above. Oil in the form of gasoline is also being burned to power automobiles, buses, trains, ships, and airplanes, with oil and natural gas also employed to heat buildings, among many other applications which include its use in manufacturing thousands of oil-based products, some of which are identified in the last chapter of this book.

For many people in the developed and developing countries, these are amenities that are part of the life they live, ones that have become essential. Yet they have come with the unintended consequence that underlies the availability of food and water slowly being compromised.

The unintended consequence of burning coal, natural gas, and oil is an overall increase of the yearly average temperature of the earth over the past 50+ years that has led to detrimental repercussions globally. The main cause of the temperature increase is the emission of heat-trapping gases into the atmosphere and the oceans from fossil fuels: coal, natural gas, and oil. The most significant of these heat-trapping gases is carbon dioxide (CO_2), whose increase in the atmosphere has been authenticated by measurements made over many decades, analogous to those that have verified the overall temperature increase.

Other possible contributors to this temperature increase, i.e., to *global warming*, such as volcanic eruptions, changed solar output, and the belching of cattle do not occur in amounts anywhere close to that produced by the burning of fossil fuels. In other words, global warming is an anthropogenic (human-caused) phenomenon, as shown later in the book.

The repercussions of anthropogenic global warming are widely acknowledged as a threat to the planet. Large numbers of young people worldwide as well as various politicians and government persons are among those who have not only grasped this but have called for concerted action to deal with it. Acceptance by politicians, however, is far from universal, some of whom persist in calling it a hoax. Not only do some politicians reject the reality of global warming, but many others, mostly ordinary citizens, have rejected it as well.

Such rejections are not new. Well over 100 years ago, the Swedish scientist Svante Arrhenius carried out a calculation which showed that doubling the amount of carbon dioxide in the atmosphere might cause global warming. His result was rejected. Unlike modern times, where the rejecters of global warming are rarely climate scientists much less scientists at all, in his day it was fellow scientists who did the rejecting. And just like in modern times, the rejections then were erroneous.

Climatology, primitive in Arrhenius' day, has now become a full-fledged science in which anthropogenic global warming is endorsed by the overwhelming majority of climatologists, whose measurements have yielded vast amounts of detailed information about earth's climate system, even as far back as millions of years ago.

So, if Arrhenius' adversaries could be brought forward in time and learn of the data that has been gathered, it is likely that they would abandon their objections and accept the reality of global warming and its causes. This is how science is practiced: access to reliable and up-to-date data will lead scientists to abandon an invalid premise and adopt the new one.

Unfortunately, this has not been the case for many of those who have rejected the reality of anthropogenic global warming and/or its harmful consequences. They persist, despite their spurious antiglobal warming claims having been consistently refuted. Some of these claims are not only recycled in blogs, but they also appear on social media, on TV and radio, and in opinion pieces and letters to the editor, especially of smaller newspapers. To ordinary readers who do not access science websites on this topic, these sources may suggest that global warming is controversial, whereas in the climate science community there is no controversy, a fact not likely to be known to the general public.

It is this recycling of bogus claims that has partly motivated my writing the book. It contains information on global warming that I have presented in the many courses on it I have taught for persons with neither a science nor a math background. Given my background as an educator and researcher, I have written the book from an educator's point of view: I want my readers to end up learning enough to understand and critically assess what is included in any sources on the subject they may encounter. This means that I have provided, either directly in the book or via reference to websites, scientific evidence underlying both the validity of global warming and its repercussions and why the claims of rejecters against it are spurious.

The book is thus a compact introduction for persons who are unfamiliar with the relevant details but want to know what anthropogenic global warming and its repercussions entail. The first chapter presents background material on the earth's climate system and some of the rejecters' claims and their refutations, including the evidence underlying the reality of anthropogenic global warming. The next chapter examines in detail many of global warming's contemporaneous deleterious consequences, while the final one considers some future scenarios and their likely consequences, as well as proposed steps to deal with them.

I mentioned above one motivation for writing the book. An equally important one is my set of personal reactions. As a scientist I am offended and insulted by the apparent refusal of so many rejecters to access and/or accept the scientific data and conclusions. Not only scientists but any rational person not gripped by an ideology should feel this way.

While these comments are based on my scientist persona, I am motivated as a human being who is deeply concerned with the planet all people live on, one that in different ways is slowly being ruined. James Hansen aptly chose Storms of My Grandchildren as the title of his excellent 2009 book on global warming and climate change. And it is my grandchildren and yours who will be affected, for all of whom we should feel a responsibility. As Chief Seathl of the Duamish Nation is alleged to have written to the US President in 1855 during the US purchase of native land, "whatever befalls the earth befalls the sons [and daughters] of earth." We must be stewards of the earth and do everything in our power to heal as much of it as is possible. My hope is that this book along with others will help motivate people to do that.

CONTENTS

INTRODUCTION

The main activities of science are collecting verifiable facts about some aspect of the universe and creating or investigating testable theories that organize and coherently explain both the existence of these facts and their relationships to one another, where a scientific theory is a well-substantiated explanation supported by confirmed facts and predictions.

This description may seem innocuous, but history has clearly shown that what it describes can be controversial. Why? Because the conclusions of science have often been seen as a threat or a challenge, for example, to certain ideologies, to religious doctrine, to authorities, to accepted wisdom or explanations, to entrenched opinion, to political aims, or to profits.

This is especially true for climate science, whose conclusions about global warming and its repercussions have been strongly denied, denigrated as false, and in many instances vehemently attacked, as have been some of the climatologists themselves. I will refer in this book to persons who have done any of this as "rejecters," who themselves have sometimes labeled the acceptors of global warming as "alarmists." As it turns out, the rejecters, whose motives are based on one or more of the threats and challenges outlined above, *have not had valid scientific evidence to support their claims.*[1] In contrast, as will be made

F. S. Levin, *Global Warming: Truth and Consequences*,
https://doi.org/10.1007/978-3-031-27023-9_0

clear in this book, the "alarmists" have had compelling reasons to sound an alarm: global warming is an existential threat to the planet.

Some of the rejecters' spurious claims and their evidence-based refutations are briefly reviewed in Chapter 1, but it should be clear from the outset that global warming[2] is real and anthropogenic, its deleterious impacts are already being felt worldwide, and its further consequences will alter life on earth as it is currently constituted. These conclusions have been endorsed by an overwhelming majority of the climatologists who have done research on it and published their results in peer-reviewed journals. In other words, these scientists have achieved *consensus* in this regard.

Despite bogus claims to the contrary, reaching consensus is a normal outcome of the scientific process. It means that the science has become settled, in that the prevailing paradigm is unlikely to change substantially. This does not mean there are no gaps to be filled in, only that the existence of any is insufficient to derail the paradigm. The filling in is sometimes done by gathering additional experimental or observational evidence, and sometimes by further theoretical analyses. This has even led in some instances to the creation and eventual adoption of a new paradigm after the original one had been shown to be either inadequate or wrong. There is, however, no indication whatsoever that the global warming paradigm, a version of which is presented in Chapter 1, is at all likely to change. So, it is worth emphasizing that global warming is real, humanity is already suffering its consequences and will continue to do so, even more deleteriously in the future unless drastic steps are taken. Whether they can be and will be is discussed in Chapter 3.

1

THE BASICS

Persistence is a characteristic of global warming rejecters, who have used spurious claims and diversionary tactics to attack the well-established global warming (GW) paradigm and its adherents. That paradigm, stated below, implies that global warming is a planetary disease, an existential one in which the earth can change deleteriously from what it is now. How this is very likely to occur is described in Chapter 3. But the purpose of this chapter is to provide the basic information needed to understand global warming and the successful efforts to describe it scientifically. I'll start with the meaning of the term global warming.

It is intimately connected with "climate," which is *weather averaged over large areas and time periods*, for instance, 30 years or more. Weather refers to atmospheric conditions such as temperature, precipitation, humidity, cloudiness, winds, and pressure, all within a localized area. The long-term averaging of weather has led to certain climate phenomena being presented as trends, for example, temperature trends.

The temperatures appropriate for assessing global warming are those that have been measured over the entire earth including the oceans and then averaged each year to yield a yearly average global temperature. The long-term trend

F. S. Levin, *Global Warming: Truth and Consequences*, https://doi.org/10.1007/978-3-031-27023-9_1

of this yearly average is highly significant: it is a measure that the climate scientists Makiko Sato and James Hansen use in their website "Updating the Climate Science" to demonstrate the occurrence of GW. Defined as a steady increase in this long-term temperature trend, it is the one I use in this book. As seen in Figure 1.3 shown later in this chapter, the long-term trend in the yearly average global temperatures has been an increasing one for decades, which means that global warming has been occurring, despite claims to the contrary.

The consensus GW paradigm that climate scientists have adopted, and rejecters have inveighed against, is more than just the claim that the trend in yearly average global temperatures has been increasing. One version of it is:

1. *Global warming has been occurring for decades and is caused by the increase of greenhouse gases in the atmosphere that trap heat radiated from the earth, thereby warming it.*
2. *The increase in atmospheric greenhouse gases is human-caused, or "anthropogenic," which means that global warming itself is anthropogenic: GW is actually AGW.*
3. *It is extremely likely that the consequences of AGW will continue to be deleterious for humanity, in both the relative short term and the long term.*

The atmospheric gases are labeled "greenhouse" because they act in the same way as the glass panes in a typical greenhouse, which are transparent to sunlight but block and trap the heat generated within the greenhouse, thereby increasing its temperature.

While the GW paradigm is the culmination of many years of scientific investigations, both experimental and theoretical, attempts to understand the role of earth's atmosphere started nearly 200 years ago, which is where this background begins.

A Little History

As far as is known, the person who first realized that the earth's atmosphere acts as a greenhouse which keeps the earth's average temperature well above freezing was the French mathematician and scientist Jean Baptiste Joseph Fourier (1768–1830). He had initially wondered why the sun's radiation striking the earth didn't continually heat it up, raising its temperature until it reached that of the sun. His answer was that the incoming sunlight is reradiated from the earth's surface as heat, i.e., as infrared radiation (IR), some of which is emitted into space, keeping the planet cooler. Using mathematics that he had created, he turned this qualitative conclusion quantitative and discovered, much to his surprise, that the earth should be far colder than it is.

It was this result that led him to conjecture in 1824 that the gases in the earth's atmosphere must act as an analog to the glass panes in a greenhouse, though he didn't know how this could happen. Fourier knew that nitrogen (N_2) and oxygen (O_2), the most abundant gases in the atmosphere, were transparent to IR; moreover, it had been assumed that this was true for all the atmospheric gases. Thirty-five years later, this assumption was tested by the English scientist John Tyndall (1820–1893). His experiments confirmed the results for N_2 and O_2, but when he tried coal gas, a flammable mixture of hydrogen (H_2) and methane (CH_4, where C is the symbol for carbon), he found it was totally opaque to IR, and a later experiment showed this to be true for carbon dioxide (CO_2) as well. While CO_2 and CH_4 are minor constituents of the atmosphere, they, and especially CO_2, play a highly significant role in global warming, as will be shown later. Tyndall knew that there was only a tiny amount of CO_2 in the atmosphere—a few parts in roughly 10,000—yet he realized (over 100 hundred years ago!) that *that was enough* to help block the heat radiated from the earth's surface, trapping and reradiating some of it back to the surface, thus keeping the earth warm.[1]

Tyndall also found that water vapor (H_2O in gaseous form) was a more important IR radiation blocker than CO_2, and he speculated that if something were to dry out the atmosphere, it could cause an ice age. The discoveries of the English geologist James Hutton and others, summarized and expanded in Charles Lyell's three volumes on geology, implied that ice ages had occurred during the earth's past, so there was much interest in trying to determine their cause(s).

Among those who tried to find a cause was the Swedish scientist Svante Arrhenius (1859–1927), who ended up discovering that doubling the amount of CO_2 in the atmosphere might cause global warming. Aware of Tyndall's speculation about drying out the atmosphere, Arrhenius initially wondered whether removing CO_2 from the atmosphere could be the mechanism that would lead to a decrease of its water vapor, which would lower the average global temperature and eventually cause an ice age. Since he could not test this by removing any of the CO_2, he did the next best and only thing: he constructed a mathematical model that tried to simulate as much of the earth's climate system as was then understood and calculated the change of the earth's temperature if atmospheric CO_2 were decreased.

His calculations were extremely laborious, done by hand, took many months, were crude, suffered from a lack of completely reliable information on how the atmosphere absorbed radiation, and omitted various climate-influencing factors, though he did include water vapor effects. He ended up with an estimate: cutting CO_2 concentrations by half would produce a drop in the average global temperature by approximately 5 °C = 9 °F (see Table 1.1 for definitions of temperature scales). This change is significant, and he thought it might be enough to bring on an ice age by means of *feedbacks*: decreasing the temperature would increase the amount of ice and snow, which would mean more of the sun's radiation would be reflected from the earth (i.e., an increase in the earth's reflection coefficient or *albedo*), which would mean less heating of the earth and its atmosphere, leading to more ice and snow, and so on to a hypothetical ice age.

This result led him to ask what would happen if the concentration of atmospheric CO_2 were doubled. After more laborious calculations, he came up with a second estimate: doubling the concentration would increase the global temperature by roughly the same amount, namely 5 °C. It was the first claim that an increase in atmospheric CO_2 (obviously a huge increase) might cause global warming based on a climate model calculation. Why "might" cause GW? Because his calculations were based on a crude mathematical simulation of the earth's climate system.

In fact, his results were rejected by skeptical scientists of his time, a rejection that could be looked on as an eerie

Table 1.1 The Fahrenheit and Celsius Temperature Scales

The freezing and boiling points of water, under normal atmospheric conditions are what sets the two temperature scales. For Fahrenheit, the freezing point of water is at 32 °F while its boiling point is set at 212 °F. Celsius, on the other hand has a much smaller range: water freezes at 0 °C and boils at 100 °C
Temperatures in each scale are related to the other by the following formulas: °C = (°F −32)/1.8 and °F = 1.8 x °C + 32, where °C and °F mean the value of the temperature in each scale that is to be converted to the other scale. Some examples are 15 °C = 59 °F, 20 °C = 68 °F, −40 °C = −40 °F

forerunner of what has occurred in our own time, except for two distinct differences: first, almost all modern rejecters of global warming have not been scientists and have made claims shown by climate scientists to be spurious; second, Arrhenius' skeptics had what were believed to be valid scientific grounds for rejecting his conclusions.

There were both theoretical and experimental grounds for these rejections. On the theory side, it was noted that Arrhenius had not included effects from other parts of the climate system, for example, winds, ocean currents, and the creation of clouds from water vapor. The behavior of these items, it was thought, would act either to modify or to prevent warming. In other words, his model was a vast oversimplification of the climate system, and therefore, its results were not valid.

A further theoretical argument was that he had ignored the alleged fact that any additional CO_2 would be absorbed by the oceans and not by the atmosphere. This allegation was an unrecognized assumption, based on the fact that the oceans contain about 50 times as much CO_2 as is in the atmosphere. Although the assumption was believed to be correct, it is not, and thus the counter-argument is also false, although it wasn't known then: the oceans do not prevent additional CO_2 from going into the atmosphere, as has recently been determined from detailed measurements, nor is the oceans' ability to absorb CO_2 without harmful effects such as coral bleaching.

A more devastating argument came from an experiment carried out in 1901 by the Swedish physicist Knut Ångström (1857–1910) who, with an assistant, measured the absorption of IR by a column of CO_2 estimated to be equal to the amount in the atmosphere from its top to its bottom. Their measurement showed that there was hardly a change in the absorption when the amount of CO_2 was reduced by one third. Since it was known that

CO_2 absorbed IR only in specific "bands" (which will be described soon) and that the amount in the atmosphere was tiny, their conclusion was that this tiny amount had saturated the bands, which meant that atmospheric CO_2 could absorb no more. Hence, atmospheric CO_2 could not trap any further heat emitted by the earth no matter how much additional CO_2 entered the atmosphere, and therefore, global warming could not occur, contrary to Arrhenius' result. The same conclusion was drawn with respect to the water vapor in the atmosphere.

The scientists who accepted this saturation argument couldn't have known it also was false, so along with it and the other objections, they unknowingly initiated a global warming controversy that has persisted to this day, except that in modern times, it is almost exclusively non-scientist rejecters who promulgate the so-called controversy, since, as noted above, the vast majority of climatologists have accepted the paradigm: for them, there is no controversy. Moreover, and in sharp contrast to Arrhenius's era, an enormous amount of pertinent information is now known about the earth's atmosphere, some of which is surveyed next.

The Major Constituents of the Earth's Climate System

The earth's climate system has four major constituents, each of which has many components. The major ones are the radiation it receives from the sun, its gaseous atmosphere, its oceans, and its land surface. The most important in the present context are the atmosphere and the sun's radiation, which brings energy to the earth. Together, they allow life to exist without huge temperature extremes (at least since the last ice age), with some of the atmosphere's constituents preventing excessive amounts of harmful radiation reaching human beings.

There are two ways of specifying the amount of energy radiated by the sun: either the total amount or the portion in each wavelength contained in the radiation. Wavelengths characterize waves, phenomena familiar from everyday experience, for example, the visible ones such as ripples in a pond or the invisible sound waves in air caused by speech, singing, or the playing of a musical instrument.

Scientifically, a wave is a periodic disturbance in a medium of some kind, such as water, air, or a musical instrument string. Periodic means it repeats, though the repetitions die out in real situations. To describe waves, scientists examine them in ideal situations, such as those with no friction and in one dimension. A standard example is a sine wave, depicted in Figure 1.1.

The horizontal line represents the undisturbed medium, while the curved portions are two full sine waves, with each of the two having one crest and one trough. The amplitude is the maximum departure from the horizontal undisturbed position, while the wavelength is the shortest distance between analogous portions of the waves.

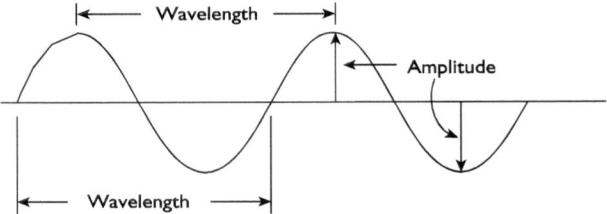

Figure 1.1 Schematic representation of a portion of a sine wave. Shown are the wavelength, which is the distance between adjacent portions of identical character (two adjacent maxima and two adjacent *nodes* or zeros), and the amplitude, which is the maximum departure from the undisturbed position

A wave with a unique wavelength is called *monochromatic*. This name comes from the study of the sun's radiation, which contains an enormous range of individual wavelengths. The range is known as the *spectrum* of electromagnetic radiation, and Table 1.2 delineates the spectrum; most of the names of its different segments should be familiar. What may not be familiar are the values of the wavelengths, which are expressed in terms of both microns and nanometers (nm). For example, the visible range is approximately 400–700 nm, which covers the six colors of the rainbow: red (at 400 nm), orange, yellow, green, blue, and violet (at 700 nm). The visible portion is what we refer to as light, and it is the association of a specific value of wavelength in the visible range with a particular color that causes any wave with a unique wavelength to be called monochromatic (one color). Examples are the light from a laser or a neon sign.

Table 1.2 The spectrum of electromagnetic radiation

Segment	Wavelengths in μm[a]	Wavelengths in nm[a]
Radio Wave	Greater than 1 million	Greater than 1 billion
Microwave	100 to 1 million	10,000 to 1 billion
Infrared	1 to 1000	1000 to 100,000
Visible	0.4 to 0.7	400 to 700
Ultraviolet	0.001 to 0.4	1 to 400
X ray	0.0000001 to 0.01	0.0001 to 10
Gamma ray	Less than 0.0001	Less than 0.1

[a]μm is the abbreviation for a micron; 1 μm = one millionth of a meter

nm is the abbreviation for a nanometer; 1 nm = one billionth of a meter

1 μm = 1000 nm

(1 meter is about 39 inches)

Human hair widths are sometimes expressed in microns:

fine hair is about 25–50 μm

medium thick hair is typically 60–90 μm,

coarse hair is greater than 100 μm, possibly as much as 180 μm

While the visible range of the spectrum is tiny, it is extremely important because the maximum energy in the sun's spectrum is in the visible, with smaller amounts mainly in the IR closest to the visible and in the ultraviolet (UV). This remains true for the radiation that reaches the earth's surface after it passes through the atmosphere, whose absorption diminishes the energy at all wavelengths, but much more in the IR and UV than in the visible. That is the reason why human eyes and those of animals have developed via natural selection to be sensitive to just the visible part of the spectrum.

Although the energies that reach the earth's surface are predominantly in the visible, almost all the energy that the earth radiates back is in the IR, as conjectured so long ago by Fourier. In terms of wavelengths, what reaches the earth's surface from the sun is predominantly short wavelength, while what is emitted by the earth is long wavelength, some of which is absorbed by the greenhouse gases and reradiated back to the earth. When radiation is absorbed by several closely spaced energy levels in an atom or molecule, it is referred to as being absorbed by a *band* of levels (see the comments above concerning saturation of CO_2).

The sun radiates a gigantic amount of energy. The *insolation* is the amount per second that reaches the top of the earth's atmosphere, and to deal with it in manageable terms, it is typically divided by the earth's surface area. This means it ends up being expressed in watts per square meter (W/m^2), where a meter is a little over 3 feet. The numerical value of the insolation at the top of the atmosphere is approximately $342\,W/m^2$, while the amount reflected from the earth's surface and from the atmosphere (clouds, gases, aerosols) is about $105\,W/m^2$, so that the amount not reflected is roughly $237\,W/m^2$. Of that, about $67\,W/m^2$ is absorbed by the atmosphere, which means that roughly 170 W/m^2, or about half the amount at the top of the atmosphere, is absorbed by the earth's surface.

The portion of the insolation that is reflected—the *albedo*—is roughly 30%. Recall that in describing how an ice age might begin, Arrhenius referred to feedback mechanisms, for instance, that lower temperatures would mean more ice and snow, which would then reflect more incoming radiation. That would increase the earth's albedo, which would mean more reflection followed by lowered temperatures, etc. In contrast to this scenario, global warming produces the opposite effect: higher temperatures lead to increased melting of ice and snow, reducing the earth's albedo, which means more of the insolation being absorbed by the earth, with corresponding increases in average global temperatures, which lead to further decreases in the albedo, etc. So, a crucial question is: has the amount of ice and snow changed in recent years? The answer is yes, as described in Chapter 2.

The atmosphere is the home of greenhouse gases, but how much of it do they constitute, and what else is there? The main ingredients in dry air (no water vapor) are nitrogen (N_2) and oxygen (O_2), whose amounts are approximately 78.08% and 20.94%, respectively. Next is argon (Ar), which comes in at about 0.93%. Taken together these three make up 99.95% of its constituents, which means that all the other components make up roughly 0.05%, among which, of course, are the greenhouse gases. CO_2 is the next most abundant at approximately 0.04%, with methane (CH_4), at nearly 0.0002%.[2]

These latter two numbers are so small that it may seem impossible that they could be causing global warming, just as rejecters have claimed. CO_2 and CH_4 play such a significant role because they are the main absorbers of IR, in contrast to the three most abundant molecules, which do not absorb it. An analog—and only an analog—for a tiny quantity of something having a huge influence is the ingesting of a miniscule aspirin to relieve a headache or an anti-allergen to combat an allergy attack. In each case, the size of the ingested object is miniscule compared to the size and weight of an adult body, just as CO_2 and the other greenhouse gases are miniscule when compared to the three, non-CO_2-absorbing main constituents of the atmosphere.

The standard unit in which greenhouse gases is measured is *parts per million by volume*, abbreviated as *ppm*. The amount of N_2 in the atmosphere is approximately 780,800 ppm, while that of Ar is about 10,000 ppm. In contrast, CO_2, which seasonally fluctuates, was nearly 415 ppm in May 2019 and roughly 410 ppm in November 2019, growing to 417 ppm in May 2020, 421 ppm in April 2021, and back to 420 in August 2022,[3] likely confounding some COVID-19 expectations (see Chapter 3). These CO_2 concentrations have grown by approximately 50% over the 1880 value of about 280 ppm, with almost all of it since 1970 being anthropogenic. And as seen in graphs on the internet, anthropogenic CO_2 ppm is the highest in the past 800,000 years.

Recall that the GW paradigm is predicated on the increase of greenhouse gases, especially CO_2, whose dramatic increase has now been quantified. The numbers themselves are clearly important, but an actual plot of the increase is much more revealing. Such graphs show that the growth has been nonlinear, i.e., is not a straight line. The nonlinear growth of atmospheric CO_2 from 1958 to August 2022 is shown in Figure 1.2. The curves with the repeated up and down fluctuations are known as "Keeling" curves, named for Charles David Keeling of the Scripps Institution of Oceanography at the University of California in San Diego.

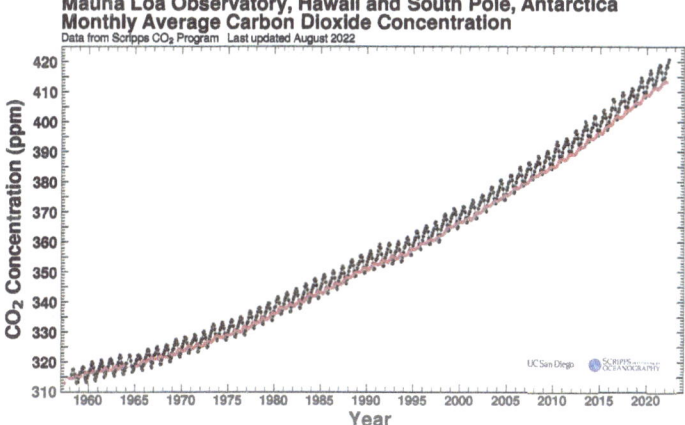

Figure 1.2 Growth of the average monthly carbon dioxide concentration. Black curve measurements are from Mauna Loa, Hawaii, the Red curve measurements are from the South Pole, Antarctica.

In 1958, he began measuring atmospheric CO_2 levels from the top of the Mauna Loa volcano in Hawaii and in Antarctica. While funding problems meant he had to abandon the Antarctic measurements, he continued the Mauna Loa ones, which his son Ralph Keeling eventually took over. The black curve is a graph of the Mauna Loa measurements, and the red one is a graph of the Antarctica measurements.[4]

The nonlinearity of these curves can be exhibited by drawing a straight line from the amounts of the red curve between 1958 and 1970. Such a line—which is linear—will lie below both curves, with the difference between them and the straight line becoming more and more pronounced. These curves substantiate the steady increase of the amount of atmospheric CO_2 that is part of the GW paradigm. This nonlinear steady increase is an exponential one, with a growth rate of 2%. The fluctuations in the Keeling curve correspond to the time of year when the measurements were carried out: there is slightly more atmospheric CO_2 in the winter when trees lose their leaves, slightly less in the summer when their leaves absorb CO_2. The maxima typically occur in late April and May, the minima late September and October. That the Antarctic curve has smaller fluctuations and lies slightly lower than the average of the Keeling curve is a result of most of the earth's vegetation being in the northern hemisphere.

The Non-dramatic Role of Water Vapor

The percentages listed above for the atmospheric constituents are for an atmosphere without water vapor. Yet water vapor is a potent greenhouse gas, whose concentration, which varies with latitude, can be as large as 30,000 ppm, an amount far greater than that of CO_2. Because water vapor absorbs IR over a much larger range of wavelengths than CO_2, shouldn't it be the cause of global warming, as has been claimed by rejecters? The answer is no, because its varied atmospheric concentrations have only increased slightly, due to the increases in CO_2 that warm the planet, thereby allowing for more evaporation and thus more water vapor in the atmosphere.

There is another way of describing this: when there is a small increase in surface temperature (here, due to increases in greenhouse gases), water vapor compensates by keeping the *relative humidity* roughly constant, while the *specific humidity* increases slightly, which means that the amount of water vapor increases slightly.[5]

In more technical language, the increase in water vapor is a feedback that responds to the increase in CO_2, not vice versa. Nevertheless, as a major absorber of IR, it must be included in simulating the earth's climate system—and it is.

The atmosphere is a complex system whose temperature and density change as one ascends higher and higher, giving it a height structure. There are five distinct layers in this temperature/density feature. Starting from the earth's surface, they are the troposphere, the stratosphere, the mesosphere, the thermosphere, and the exosphere. The height of each layer above the earth varies with latitude and to some extent with the season. For the purposes of this book, the troposphere is the most significant. Its height typically varies from about 7 to 20 km (or about 4–12 miles, where 1 km = 1000 meters). Its average temperature decreases from about 14 °C near the earth's surface to about –60 °C, after which it remains constant to the top of the troposphere, where the stratosphere begins, and the temperature starts to increase. (The portion where the temperature is constant is the tropopause.) The troposphere contains about 70–75% of the mass of the atmosphere, roughly 99% of its water vapor, and almost all the clouds, apart from those above the two polar regions (See note 2). The earth's weather occurs in the troposphere, which is significant for GW not only because it must be included in global climate models[6] (GCM) that simulate the earth's climate system, but also because of a verified prediction about it that was made from a GCM, as described below.

Anthropogenic CO_2

The GW paradigm states that human activities are the cause—the source—of the increase in atmospheric greenhouse gases, especially CO_2. The vast majority—nearly 80%—of these human activities are the burning of fossil fuels, with most of the remaining increases coming from industries such as cement manufacturers and from the deliberate reduction of forests. Is there actual proof that these activities are the source? The answer is yes. It is based on an isotope[7] of carbon. Normally occurring carbon—carbon 12, the one found in just about everything from molecules in the body to sugar to fossil fuels—makes up about 99% of its three isotopes. The other two are carbon 13 and carbon 14, each a little heavier than the preceding one. All three are found in the atmosphere and in fossil fuels and plants. But fossil fuels are derived from ancient plants, which like ordinary plants have the ratio (carbon 13/carbon 12) that is about 2% lower than in the normal atmosphere. The burning of fossil fuels, which puts CO_2 in the atmosphere, should therefore lead to a decreasing value of this ratio in the atmosphere—and measurements have verified this. Conclusion: the increase of atmospheric CO_2 is indeed anthropogenic.

From this survey of the earth's climate system, I turn next to a review of some of the rejecter claims and their refutations, which involve pictorial proofs of global warming and the accuracy of global climate model calculations in both predicting and postdicting.

Representative Rejecter Tactics and Arguments

The negative responses to the global warming paradigm fall into two broad categories: doubts and denials. The purpose of the first one is to create uncertainties about global warming among persons and entities, especially those unfamiliar with the paradigm. Denials, on the other hand, usually focus on specifics, and these are the only responses considered here.[8]

Rejecters of the global warming paradigm have presented a great variety of counter arguments and attacks against it. I will discuss only a very small subset of them, since even a small subset is sufficiently illustrative. A selection of them is listed in Table 1.3.

All are not only false but have been refuted.[9] Refutations of them include citing data that was ignored or not accessed, quoting the full statement of a cherry-picked claim, showing how some accusations completely misrepresent the scientific evidence, and using well-established conclusions to show that a claim is false. Those in the table have been selected in part so that

Table 1.3 Some Counter Arguments to the Consensus Paradigm of Global Warming

1. There is no consensus
2. Average global temperature increases are within normal variations, which means there is no global warming
3. As many glaciers are growing as retreating and so this indicator of global warming is false
4. Average global temperatures are cooler than in the past, so there is no global warming
5. Weather is a chaotic phenomenon, so global climate predictions are meaningless
6. Since global climate models cannot hindcast (or postdict) accurately, their forecasts cannot be trusted
7. Eons ago, temperature increases occurred before CO_2 increases, therefore the opposite claim that increases in CO_2 are causing temperatures to increase is erroneous: CO_2 does not cause global warming
8. Global warming is occurring, but its effects will be small and beneficial to humankind
9. The sun, not human activities, is causing temperature increases
10. As anthropogenic CO_2 is a small fraction of all CO_2 increases, human activities cannot cause global warming
11. There has been no global warming since 1998
12. Antarctica is gaining not losing ice

readers can see the particulars and range of some of the false claims, some recent, some not.

In discussing the rejections and refutations, it is useful to introduce the symbols used for the atmospheric concentration of CO_2, for temperatures, and for their increases. The concentration of carbon dioxide will be indicated by the symbol $[CO_2]$; T_{av} will specify the yearly average global temperature, while yearly average global temperature changes from a specific baseline—referred to as *anomalies*—will be denoted by ΔT, where Δ is the capital Greek letter delta, often used to indicate changes in a quantity.

I'll start by recounting a few that I've been personally involved in that illustrate rejecter tactics. The first instance comes from a scientific society's Newsletter whose co-editor for the first time in the Newsletter's history wrote editorials and letters that espoused an armful of rejecters' claims. Of the many, I've selected the following two. One is that Al Gore supposedly stated in "An Inconvenient Truth" that sea level rise would be 20 feet by 2100, but this is egregious cherry picking, because what Gore wrote is that *if* Greenland melted or broke up, or *if* half of it and a portion of Antarctica melted or broke up, then sea levels would rise 18–20 feet by 2100. The second of these spurious claims was that "science is about facts … not consensus" (the omitted words in the ellipses … are not part of that claim). My rebuttals of these and many of the other bogus claims were published in a subsequent issue of the Newsletter.

I have also written opinion pieces and letters to the editor in local newspapers where I have lived that refuted various rejecters' claims. One claim was that there has been no global warming since 1998. This is debunked below. The strangest claim was that we are in a transition period of the Pleistocene, which will lead to colder future conditions than now. Unfortunately for the writer, who obviously believed it, the premise of this claim is totally fictitious: the Pleistocene era ended about 11,700 years ago.

A repeated claim in various venues has been that no proof exists that human activities are the dominant cause of the warming of the earth's atmosphere over the past 50 years—at least as proof is understood in science. The actual existence of one, as cited in my responses and delineated above, refutes this claim. It seems likely that rejecters have either not wanted to learn if a proof exists or have not been able to understand it. Instead of anthropogenic activities as the cause of atmospheric temperature increases—mainly fossil fuel burning—rejecters have claimed it to be galactic cosmic rays, or increased solar radiation, or volcanic emissions of CO_2, or methane emissions from belching cattle. Measurements have shown that none are possible causes.

This smattering of false claims and their refutations should help establish in readers' minds that even minor venues like local newspapers can be used to present specious arguments. It is unknown how often they have gone unchallenged, which makes this kind of tactic so guileful.

I now turn next to examples involving persons well known in the climatology and rejecter communities, though some may not be familiar to all readers of this book. They have been important in the climate wars, further details of which can be found in some of the sources listed in the bibliography and on the internet.

The first concerns the 1988 US Senate testimony of James Hansen, then director of NASA's Goddard Institute of Space Sciences. A rejecter's follow-up 10 years later employed an often-used tactic: the cherry picking of information—data and/or calculations—to make his false claims seem reasonable if not damning. It entailed what one newspaper columnist labeled as fraud.

Included in Hanson's testimony was a graph in which the annual average global temperature anomaly ΔT was plotted against time,[10] beginning with temperature data from slightly before 1960 and ending in 1988. He compared this data with the ΔT results from three global climate model calculations,[11] each of which was based on different assumptions for the growth of CO_2: in scenario A, the growth was assumed to be exponential (a growth analogous to that of CO_2 in the Keeling curve); in scenario B, the growth was linear, i.e., it increased by the same amount every year—a business-as-usual hypothesis that he thought was the most plausible; while in scenario C, there was no growth, just the same amount year after year.

However, the results of the calculations did not end in 1988: they were extended to 2019 and thus were forecasts. Though the results were in reasonable agreement with each other over the period 1960–1988, none of the three fit the data very well. More importantly, the forecast difference in the ΔTs from scenario A compared to those from both B and C became large by 1995, while the ΔTs from B and C did not begin to diverge noticeably until after 2005.

Hansen's testimony had a large impact. First, he told the Senate Committee that it was 99% certain that the warming trend was not natural but was caused by the buildup of CO_2 and other gases in the atmosphere. And in the interview with reporters after the hearing, he stated that "It's time to stop waffling so much and say that the evidence is pretty strong that the greenhouse effect is here." These strong statements were picked up by the national media.

By 1988, most climate scientists had concluded that GW—Hansen's greenhouse effect—was a reality caused by human activities and would continue for some time until CO_2 emissions could be drastically reduced, though they did not have strong evidence for its being anthropogenic. Nonetheless, Hansen had made three different ΔT forecasts, so it was of interest to learn which, if any, was reasonably accurate. It is in this context that the cherry-picking enters.

In 1998, what was claimed to be Hansen's original graph was shown in US congressional testimony by Patrick Michaels, a former editor of an anti-GW magazine, except that the graph Michaels displayed had been cherry-picked: it had only scenario A's forecast on it (those of B and C had been excised). This graph showed that over the ten-year period 1988–1997 Hansen's apparent forecast was for ΔT to increase by 0.45 °C, whereas the measured value (the datum) was approximately 0.11 °C! Therefore, he claimed, Hansen was wrong by 300% and so climate models could not be trusted, which some rejecters had alleged all along. However, if Hansen's entire graph been presented, it would have shown that the ΔTs from scenarios B and C agreed reasonably well with the 0.11 °C value. In an unsurprising development, Michael's spurious claim was celebrated in the rejecter universe.[12]

The next example of rejecter counterarguments involved what are known as *proxies*. Proxies are entities from which temperatures can be deduced. To understand why they are needed, recall that global warming means that the multiyear trend of the yearly average global temperatures T_{av} is increasing. To determine the trend, one must know the temperatures. Nowadays that means using temperatures provided by thermometers around the globe and averaging them. Suppose, however, that one wants to determine temperatures going back in time, perhaps very far back: how does one do so when reliable thermometers only go back to approximately 1880? The answer is via proxies.

One such proxy is oxygen isotopes in ice cores. Using them involves measuring both the small amount of oxygen with 8 protons and 10 neutrons in its nucleus and the amount of the much more abundant and normal oxygen isotope with 8 protons and 8 neutrons in its nucleus. The ratio of these two isotopes is related to the temperature when the different portions of the ice core were formed. As it turns out, both the time of formation and the amount of CO_2 at that time can also be obtained. From these data, a graph of how temperature and CO_2 varied over time can be constructed. Such graphs, some going back as much as 800,000 years, have been created using ice cores extracted from Greenland and Antarctica, and they show a strong correlation between the time variation of the two quantities: the maxima in each are close in time, with the peak in CO_2 preceding that of the temperature in most instances, although there are exceptions where the temperature peaked before CO_2, a point addressed below.

In the late 1990s, climate scientists Michael Mann, Raymond Bradley, and Malcolm Hughes (MBH) began using proxies to obtain temperatures going back hundreds of years. Their main ones were ice cores, tree rings, and corals, each valid for different parts of the earth. From these, using sophisticated statistical procedures, they deduced northern hemisphere temperatures going back to 1400. Their article, published in 1998 (MBH98), contained a graph of their results, plotted as the anomalies ΔT relative to the average or "mean" value of the temperatures over the interval 1902–1980. There were press releases about it that led to a great deal of media attention. While there were large uncertainties in the deduced ΔT data, it was clear that starting around 1970, the anomalies displayed a maximum whose peak value exceeded that of any of the preceding ones by about 0.2 °C.

This MBH98 result suggested that GW was occurring, but it was not conclusive evidence, and MBH did not claim it was. Nevertheless, the result was attacked by rejecters. A false claim by one GW rejecter was that MBH had stopped at 1400 to avoid displaying the temperatures of the so-called medieval warm period,[13] the maximum of which was supposedly higher than any of the ones in MBH98.

If this latter supposition was true, it meant that GW had not occurred. However, this claim was erroneous. Why? For one, MBH had stopped at 1400 because their proxies were not reliable enough to provide sufficiently accurate temperature data before then. In other words, it was a scientific reason that had nothing to do with the medieval warm period. Secondly, the temperature curves for the medieval warm period were approximate estimates, drawn by hand to illustrate what the temperatures might have been: they were not claimed to be accurate. A further point was that the medieval warm period was essentially limited to England and not the whole northern hemisphere, in contrast to the MBH98 results. Because the medieval warm period curves appeared in the IPCC's first Assessment Report, published in 1990,[14] it gave them a degree of credibility and the rejecters a further reason to attack MBH98.

Soon after their 1998 publication, MBH decided to try to look for other proxies they thought would be valid to extend the time frame over the northern hemisphere back to the year 1000. They succeeded, publishing their new results in 1999 in an article later referred to as MBH99. It included a graph from 1000 to 1999 that displayed the ΔT anomalies as departures from the average value of the temperature over the 30-year period 1961–1990.

Two features from the MBH99 article are significant. First, they found that the 1990s was the warmest decade over the whole thousand-year period, with the maximum anomaly occurring in 1998. In other words, 1998 was the warmest year for the 1000 years they analyzed.

Second, while their so-called hockey stick graph did display a medieval warm period from about 1000 to about 1200, that period's maximum peak was *less* than the average temperature over the 30-year period 1961–1990. In other words, there was no question that any temperatures of the medieval warm period were equal to those of modern times, thus quashing the anti-GW claim.

There is only one other aspect to this story that I will comment on, for it involves the professional lives of Michael Mann and his English colleague Phil Jones, who was the director of the Climate Research Unit (CRU) of England's University of East Anglia.[15] Mann and Jones were attacked in 2009 when emails stored in the CRU computers were hacked and made public. The resulting furor became known as "Climategate" after rejecters claimed that the emails were a "smoking gun" that showed the MBH98 temperature reconstructions were fraudulent. It was, they trumpeted, the final nail in the coffin of global warming. They stated that these emails revealed both the fabrication and hiding of data that showed the earth was warming, that climatologists had subverted the peer-review process, and Jones had evaded requests for data.

As it turned out, the rejecters' claims about the hacked emails were all shown to be false, while Mann and Jones were exonerated by various panels of non-scientists in the USA and in England that were convened to investigate these claims. In particular, the claim that data had not been properly shared was completely undercut when it was revealed that Jones had been required not to share certain data that had been provided exclusively to him. And finally, in what seems to me make the rejecters attacks truly nonsensical is that the hacked emails were from 1999 but were claimed by the rejecters to refer to the 2009 data!

The foregoing example identifies issues over which particular climatologists have been attacked, and why the attacks failed. In contrast, the next example concerns the so-called GW "hiatus" or "pause" that led different reject-ers to claim that starting in 1999, there had been no global warming for either 15 or 18 years. Inaccurate and/ or insufficient data initially supported the claims. In the 15-year case, there was a lack of sufficient data from the Arctic as well as biased data from ocean measurements of temperature. When these were addressed by re-evalu-ating and adding to the Arctic data along with the biases being corrected, it became evident that the revised T_{av} data refuted the 15-year hiatus claim.[16]

The 15-year claim involved both data that needed improving and rejecters who were not scientists. In con-trast, the 18-year claim concerned faulty data that was consistently endorsed by three climatologists as well as rejecters. It was a claim that ended up being the subject of a US Senate hearing. The faulty data in this case came from a series of satellite-derived measurements from 1999 to 2015 that showed much lower values of T_{av} compared to that of 1998. This supposedly proved that the earth had not warmed since 1998, and therefore, that global warm-ing was a fiction: i.e., those who had denied the reality of GW had been right all along. Of course, it did no such thing, as noted above.

There have been two major sources for the satellite data: from the University of Alabama at Huntsville (UAH) and from Remote Sensing Systems (RSS). How the errors in the UAH satellite data had occurred and the touting of these results as the best data available by three climatolo-gists as well as Senator Ted Cruz as part of his US Senate hearing are the subject of a short video titled "Retrieval Algorithm."[17]

As described in the video, it is particularly telling that satellites measure radiation and not temperatures. Two additional steps are required to obtain the temperatures:

first, voltages are deduced from the radiation information, and then from them, temperatures are obtained using a "Retrieval Algorithm." The video shows one of the temperature sets derived from the RSS satellite data, the graph of which was displayed at the hearing by Senator Cruz, who used it to claim that there had been no warming for 18 years. His claim relied on the fact that all the T_{av} values which occurred after the big T_{av} spike in 1998 (i.e., from 1999 on) were smaller than it. However, the more reliable surface-based T_{av} values clearly display warming—and no 18-year hiatus.

Not discussed in the video is RSS scientist Carl Mears' statement from another video that the uncertainties in the satellite-derived data are five times greater than those from surface-based temperatures, which could explain why the graph based on his temperatures does not show the same warming that is obtained from surface-based temperatures. Furthermore, Mears has also noted that a stronger case can be made for using surface temperature datasets, which he considered to be more reliable than satellite datasets; the former ones agree with each other better than the various satellite datasets do.

Global warming is caused by the decades-long increase of atmospheric CO_2 and the much smaller amounts of other greenhouse gases, which arise mainly from the burning of fossil fuels: coal, oil, and natural gas, plus a small portion coming from production of cement. Globally, these sources are responsible for nearly 80% of the greenhouse gas emissions. Table 1.4 shows the global percentages that come from the different fuels and cement in 2020. As stated in the Climate Central website, the sectors in the USA from which greenhouse gases were emitted in 2020 are electricity generation (28%); transportation (29%); industrial, commercial, and residential (34%); and agriculture (9%). These numbers vary widely worldwide.

Table 1.4 Global Fossil Fuel Sources and Percentages of Atmospheric CO_2 in 2020

Source	Percentage (%)
Coal	40.2
Oil	33.3
Natural gas	21.7
Cement	4

Since the increases in CO_2 are anthropogenic and the paradigm states that these increases cause global warming, then proof that GW is real will validate the paradigm. The graph in the next section showing how T_{av} has changed over time is the desired proof. Examples illustrating how well the results of global climate model (GCM) calculations agree with the measured values of T_{av} are also included in the following section. This latter topic is important because these models have been used to make projections of the likely consequences of further increases in GW, projections that are reviewed in Chapter 3.

Before turning to the behavior of T_{av} over time and the GCM calculations, I will consider one additional false claim: as can be seen in graphs of CO_2 versus temperature that go back to roughly 400,000 years ago, the rise in temperature during some intervals preceded the rise of CO_2, which has been alleged to mean that nowadays global warming has not been occurring. This allegation is countered by the rapid increase in recent times of anthropogenic CO_2 as compared to paleolithic times, rapid increases that occurred before the corresponding rise in temperatures, and not vice versa. It is also the most rapid for a long period of time.

So, why is the above claim false? The answer is related to the occurrence of ice ages and their subsequent interglacial periods. The basic theory underlying their occurrence was put forward in the 1920s by the Serbian scientist

Milutin Milankovic. Ice ages warm up and become inter-glacials when the elements of a Milankovic cycle[18] cause more radiation to heat the earth; that heat helps to release CO_2 locked up in the ice. In other words, the end of a Milankovic cycle causes CO_2 increases to lag temperature increases, with such lags typically lasting about 800 to a thousand years. And once enough time has passed and there is sufficient CO_2 in the atmosphere, this increase in CO_2 causes further increases in earth's temperature, just as is happening now: the increases in temperature have come after and not before those of CO_2. The ancient lag in the rise of $[CO_2]$ due to Milankovic cycles was predicted in 1990 but verified only later. To reiterate: the fact that CO_2 and temperature increases in the past were not completely synchronous with the current situation in no way means global warming is not happening now. That it is happen-ing faster than at any time in the recent past is a further sign that it is anthropogenic.

Yearly Average Global Temperature Anomalies

The behavior of the average yearly global temperature anomalies ΔT over the period 1880 to 2022 is presented in Figure 1.3. Probably, the first thing that jumps out at you from this figure is how "jumpy" it is: the values fluc-tuate a great deal. The black squares are the yearly global temperature anomalies ΔT, measured as the deviations from the average of the temperature values between 1880 and 1920, which is approximately 13.7 °C. The 12-month running-mean and the 132-month running-mean curves are the results of averaging the ΔTs in two different ways (where "mean" in the figure means "average"). The straight

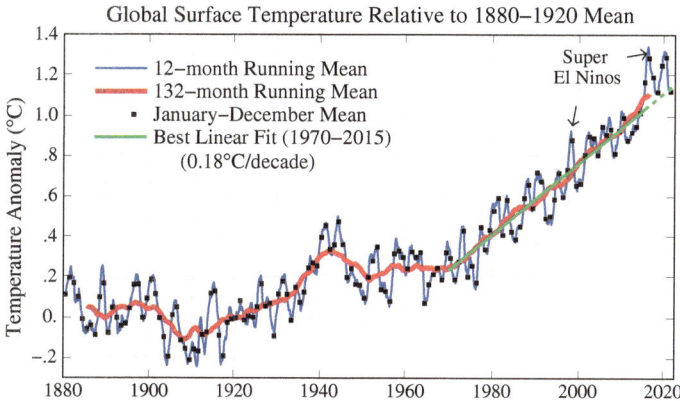

Figure 1.3 Global warming temperature anomalies over the period 1880–January 2022.

line in green—the best linear fit—is the *trend* of the temperature anomalies between 1970 and 2021. The behavior of this long-term trend is the previously noted measure used by Sato and Hansen to demonstrate the existence of global warming. The linear increase of the trend provides the proof that global warming has been and is occurring.[19]

Given the steady increase of $[CO_2]$ seen in Figure 1.2, it is no surprise that T_{av} should also increase, but what is striking in this regard is how differently these two increases have occurred. That of $[CO_2]$ is an exponential whose fluctuations are uniform, making $[CO_2]$ consistently larger in winter and smaller in summer, each by small amounts. In contrast, the "jumpy" fluctuations in ΔT are not only not uniform they undergo sharp spikes and dips, as highlighted by the blue line of the 12-month average.

Processes that have contributed to the ΔT fluctuations include large amounts of aerosols emitted by volcanic eruptions, and the phenomenon called "El Niño Southern Oscillation," or ENSO, which occurs irregularly in the

Pacific Ocean at roughly two-year to seven-year intervals, and its opposite La Niña. I'll consider El Niño first.[20]

In the normal situation, equatorial winds (the Pacific trade winds) push warm water to the west, while the cold water of the Humboldt current upwells near the northwest coast of South America, where fishermen rely on it to help increase their catch. In an El Niño occurrence, west winds drive unusually warm waters to the northern coast of South America, overriding the upwelling cold waters. These warm waters not only disrupt fishing they can also add heat to the atmosphere, thereby increasing T_{av} without any extra CO_2 being present. This occurred so dramatically in 1998 and 2016 that the ΔT spikes in Figure 1.3 have each been designated as a "Super" El Niño. Once an El Niño subsides, which can take a year or more, the atmosphere loses the extra heat, whereupon T_{av} decreases for the following year(s). This is precisely what is seen in the 1999–2001 values of ΔT as well as those for 2017 and 2018. Note that the former declines are over 3 years, not the 18 once claimed by rejecters. Note also the ΔT value for 2019: it is not only the second highest, but it also occurred without a strong El Niño and is thus a vivid demonstration of the reality of GW. Furthermore, the 2020 and 2016 values are the same, making them tied for the highest yet observed, with that of 2020 occurring despite a partial La Niña, and likely confounding a COVID-19 expectation. For a full perspective, Table 1.5 lists the eleven hottest years globally.

As implied in the foregoing, in a La Niña there is a big increase in the cold water in the eastern Pacific, which can act to draw heat from the atmosphere. This activity can work in concert with the loss of the El Niño-induced heat, leading to lower values of ΔT. It has done this numerous times, even partially offsetting the heat from an increase in CO_2.

Table 1.5 The Eleven Hottest Years Globally as of 2022[a]

ΔT Anomalies Relative to the 1880–1920 Mean[b]		
1	2020	1.29 (1.292)
1	2016	1.29 (1.287)
3	2019	1.25
4	2017	1.19
5	2015	1.17
6	2021	1.12 (1.121)
6	2018	1.12 (1.120)
8	2014	1.02
9	2010	0.99
10	2005	0.95 (0.948)
10	2013	0.95 (0.947)

[a]Data as of 2022
[b]Temperatures in °C

Turning now to aerosols, when enough enters the stratosphere, they will lead to lower T_{av} values, but in a very different way from La Niña. Sufficient amounts of aerosols can achieve this by reflecting sunlight, thus decreasing the total amount of solar energy that reaches and heats the earth's surface. Recent examples of aerosols emitted in volcanic eruptions (sulfur dioxide—SO_2—which gets converted to sulfate—SO_4—compounds in the stratosphere) are Mt. Agung in 1963, El Chichon in 1982, and Mt. Pinatubo in 1992. The Mt. Pinatubo eruption is of interest for two reasons. First, it not only occurred during a prolonged El Niño, but overwhelmed it: the aerosols remained in the stratosphere for two years, reflecting sunlight and noticeably lowering T_{av}. Second, because the quantity of aerosols emitted was determined, it is a quantity that was used in successful global climate model calculations, as described in the next section, as was the case for those of 1963 and 1982.

Volcanic eruptions also emit CO_2 that enters the atmosphere. As noted earlier, rejecters had erroneously claimed that CO_2 from volcanic emissions overwhelmed those from fossil fuel burning, so that anthropogenic contributions could not cause GW. Mt. Pinatubo is an example that shows why the rejecters have been wrong: it emitted about 42 million tonnes of CO_2 in 1991, whereas the total anthropogenic emissions then were approximately 23 billion tonnes, nearly 550 times as much as from Mt. Pinatubo! Case closed.

The reality and extent of global warming is based on measurements of the type shown in Figure 1.3; its jumpiness is not a negative factor, since the key is the linear trend. Nevertheless, the sharp dips and spikes of such graphs have not only been fodder for rejecters but have led some climatologists to propose an alternative measure of GW, one that does not exhibit such dramatic fluctuations. The alternative is the amount of heat going into the oceans. Measurements of this heat, especially in the upper 2000 meters, have not only shown an increase; as noted above, it has been accelerating, just like T_{av}. A graph of its yearly averages was published in January 2022 by an international collaboration of scientists from China, Italy, and the USA. They found that the highest annual average ocean heat content in the top 2000 meters occurred in the past three years, with each amount larger than the preceding one. The fluctuations in these yearly ocean averages have been much smaller than those of Figure 1.3, while the overall acceleration of the amount of heat going into the oceans is clearly seen in the graph. In other words, measuring the amount of heat going into the ocean might be a more straightforward indicator of global warming, as had been proposed in connection with the claim of a 15-year GW hiatus.

Even so, the information in Figure 1.3 remains crucial. As of 2017, the years 1998, 2002, and 2006 were tied as the least warm of the top 10 warmest years (relative to the 1880–1920 average). And then came 2018, whose ΔT pushed the preceding three out of the top 10. In other words, farewell to 1998 as the once-touted reason for claiming a GW hiatus. It is telling that the 11 years from 2005 on have been the warmest, with the heat in both 2020 and 2021 having occurred with an accompanying La Niña, as was also the case for the amount of ocean heat, thereby underscoring the reality of anthropogenic global warming.

While average yearly global temperatures are the scientific evidence used to determine if global warming is real, such global averages do not tell the whole story: the warming is not uniform across the earth. For example, the Arctic is warming much faster than most of the earth, with the temperature there in August 2019 hitting 34.8 °C, while it was reported in June 2022 that areas of the Barents Sea were warming 7 times greater than the global average! The rate of warming in the Tibetan Plateau is estimated to be up to three times faster than the global average, which is also being exceeded by the warming of the northwestern peninsula of Antarctica, which experienced a temperature of 18.3 °C in February 2020. If these faster rates continue, the long-term consequences will be devastating globally and not just for these regions themselves, as explored in the next chapter.

Global Climate Models and Temperature Data

With the warming trend verified, an obvious question is how well global climate models simulate the behavior of the measured temperatures. Out of a myriad of possibilities, I have chosen two examples; they illustrate how GCMs are used in data analyses. Before getting to their details, listed in Table 1.6 are some of their many successful predictions.

Table 1.6 Successful Predictions/Hindcasts of Global Climate Models

1. That the earth would warm and about how fast, and about how much
2. That the troposphere would warm and the stratosphere would cool
3. That winter temperatures would warm more than summer ones
4. Greater temperature increases as one moves poleward
5. That the Arctic would warm faster than the Antarctic
6. Hindcasting the magnitude and duration of the cooling from the Mt. Pinatubo eruption: $-0.3\ °C$ and 2 years, a result in which the model's prior and post temperatures were in good agreement with the measured values
7. Hindcasting the last glacial's maximum sea surface temperatures, which disagreed with the paleo data until improved paleo data ended up agreeing with models
8. The response of southern winds to the ozone hole
9. The poleward movement of storm tracks
10. The rising of the tropopause and the effective radiating altitude
11. The clear-sky super greenhouse effect due to increased water vapor in the tropics
12. The near constancy of relative humidity on a global average
13. Strong warming in the upper troposphere, the "tropospheric hotspot"

The table establishes that global climate models can hindcast accurately (as will soon be shown visually). It is noteworthy that the initial disagreement (item 7) between the paleo (ancient) data and the prediction was resolved *when the data was improved.* This is reminiscent of the warming-hiatus claims that relied on the satellite data which was later shown to be faulty. Not that this type of disagreement happens with some frequency: one just needs to be careful when considering claims that a GCM calculation is false. As far as Table 1.6 is concerned, its results show that the GCM codes can provide worthwhile simulations. Like many computer codes, they have short-comings, but these do not interfere with their ability to produce very reasonable results.

At the very least, partial success of GCMs in fitting the measured values of average yearly global temperatures over time underscores their utility, even if there is not complete agreement between the measured and the calculated temperatures. For instance, in attempting to reproduce temperatures over a limited range, the calculated values may display the type of fluctuations seen in the data, but not always at the same time nor with fluctuations as great. However, and it is a very important however, if the long-term trend is closely reproduced, the use of the GCM is validated.

With these introductory comments, I return to the two examples mentioned at the beginning of this section. Given that one of them deals with paleo data, it may be a surprise that theory in this case hindcasts unexpectedly well, although it is an indirect outcome of a GCM, namely the result of a theoretical procedure based on standard climatology and GCMs. The other example shows how different starting conditions lead to slightly different simulations which get the peaks and valleys in approximately the same positions as the data, though they do not always agree perfectly with them, as noted previously.

I'll consider the non-paleo example first. The GCM that produces the results displayed in Figure 1.4 included in its input entities known as "forcings,"[21] which help to simulate earth's complex climate system. However, neither the measured temperatures nor their trend is an input to the computer code: they are the data that the calculated outputs of the code are to be compared with.

With this background, let us now examine some of the details of Figure 1.4, whose five computer runs are based

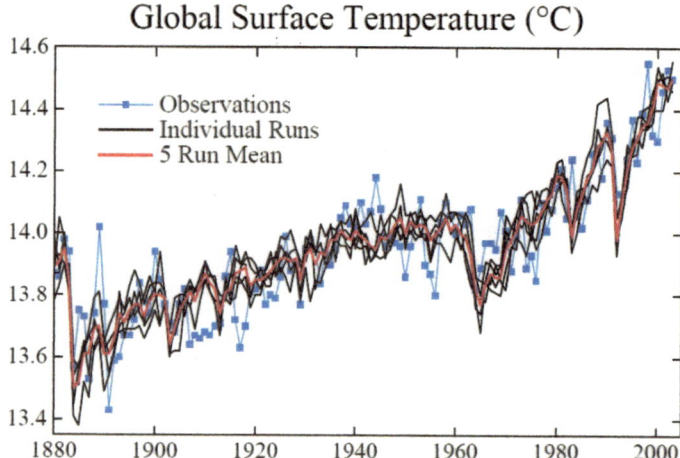

Figure 1.4 Global average surface temperatures from 1880 to 2004, the period for which the GCM calculations, in red and black, were done. The comparison with the data, in blue, is reasonably good, with most hills and valleys lining up. Hindcasting obviously works.

on different choices of the initial oceanic and atmospheric conditions. The calculated results are the black lines, some of which are so close that they cannot be distinguished from one another. The average of the five runs is indicated by the red line labeled "5 run mean." All these results yield very acceptable agreement with the data, in that the positions of the fluctuations tend to agree, even though the values of some of them are not all accounted for. In particular, the overall increase of the data from roughly 1970 on is well approximated by the GCM calculations, which clearly indicate the occurrence of global warming. Despite their shortcomings, these GCM results overall are reasonable; in addition they hindcast well enough to show that a GCM is capable of this.

As a final point, note the striking dips for 1963, 1982, and 1991 in the GCM runs. These are the years of the volcanic eruptions that produced significant amounts of aerosols in the stratosphere. The reason that these dips show up in the calculations is that the forcings included the amounts of aerosols produced by the eruptions in those years.

The results in Figure 1.4 are typical of those obtained from many other GCM calculations: reasonable but not necessarily perfect agreement with the data, whose over-all features are usually obtained well enough, for example, fluctuations and trends, as in the present example. It is typically the case that different GCM calculations which try to fit the same data do not always agree with each other (as seen for instance in various IPCC Reports). This occurs because each of the calculations is based on a different code and/or set of inputs. Yet, as noted below, that the temperatures predicted by the output of the various computer codes used in the IPCC's 2007 Report are statistically indistinguishable from one another and from the data is testimony to the accuracy of those simulations and the ability of the scientists who created them. (The GCM that generated Figure 1.4 was from the Goddard Institute for Space Studies in 2004. Newer GCMs are much more sophisticated mathematically, yet even one as "old" as that used to create Figure 1.4 was able to produce good results.)

I now turn to the paleo data example mentioned above. In contrast to the relatively narrow time period 1880–2004 of the preceding analysis, Figure 1.5 covers the much larger time period of 400,000 years. The upper curves in the graph are the climate forcings induced by both greenhouse gases and ice sheet variations. It is from these forcings that temperatures are calculated using the theoretical GCM-based relation between forcings and temperature indicated previously. The lower pair of curves compares the calculated temperatures (in red) with the

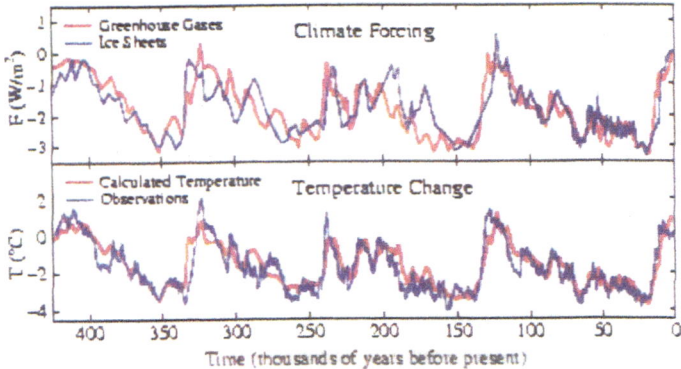

Figure 1.5 Upper curves are the two forcings that led to the calculated temperatures in red in the lower panel. They are compared with the observations (measurements) in blue, also in the lower panel. The agreement is very good over the entire 400,000-year range.

observations (in blue) obtained from ice cores, as discussed above. The agreement between these two lower curves is almost uncannily good, although the 400,000-year time period means that only gross features are captured. It is an impressive example of a type of hindcasting that GCMs can produce.

A very important corroboration of the accuracy of GCMs has been published recently: all the GCM *predictions* from the 1970s through the ones appearing in the 2007 IPCC Report have been compared with the actual measurements. As noted above and with slight caveats related to the forcings, 14 of the 17 results *were found to be statistically indistinguishable from the observations*—and the other 3 are not far off.[22] In other words, GCMs are sufficiently reliable that their forecasts should carefully be considered.

GCMs are used in the IPCC Reports to make what are called "projections" (rather than predictions) for future repercussions of global warming, a topic explored in Chapter 3. But one need not go to them to know what kinds of consequences can and very likely will occur from global warming and fossil fuel extraction because some have already occurred. Among them are increases in sea levels, decreases of ice, various effects on the earth's land masses, the inducing of earthquakes, and oil spills. These repercussions, examined in the next chapter, are already posing problems, ones that will become worse, and very likely much worse.

I will close this chapter by noting that proving an argument to be false has not prevented some rejecters from claiming otherwise. In contrast, there is a consensus among almost all climate scientists around the world that global warming has been occurring and that it is anthropogenic. Its reality and the likelihood of its deleterious future effects have been acknowledged by the US Department of Defense, 13 US Federal departments and agencies, a consortium of clergymen, some US politicians, many scientific organizations worldwide, the United Nations, the World Meteorological Organization, the World Bank, and the International Monetary Fund, as well as the governments of many countries. Unfortunately, even though more than half of the US population accepts its reality and its destructive consequences, the previous Republican majorities in the US Senate had not responded to it. Indeed, many US Republican politicians do not publicly accept the reality of global warming and seemingly have no interest in learning the factual basis for it. Consequently, no effective action had been taken in the USA as of the time of writing to try to mitigate its effects, though attempts are under consideration from the current US Democratic administration and Congress, as discussed in Chapter 3. Whether they will succeed remains to be seen.

2

CONTEMPORARY REPERCUSSIONS OF GLOBAL WARMING

This chapter describes some of the wide ranging and dele-terious effects of GW that have occurred as of the time of writing and some of the negative consequences of extract-ing and transporting fossil fuels. It thus transitions from denials and refutations to the detrimental repercussions of global warming, the planetary disease from which the earth is unlikely to fully recover, if it does at all, given the forces of greed and impotence that characterize our time, the Anthropocene.[1]

Sea Level Rise

When ocean water is heated, it undergoes thermal expan-sion. As there is no room to expand sideways or down, the expansion occurs by raising the level of the ocean, known as sea level rise. Global warming causes sea level rise in two different ways: via the additional heat that goes into the oceans and by water from melting ice sheets and glaciers.

Since oceans are found everywhere on the earth, a broad-brush way of characterizing sea level rise is to average

© The Author(s), under exclusive license to Springer Nature
Switzerland AG 2023
F. S. Levin, *Global Warming: Truth and Consequences*,
https://doi.org/10.1007/978-3-031-27023-9_2

the values from the various parts of the globe, analogous to averaging yearly global temperatures to obtain the trend. While the averages obtained by different sets of measurements can and sometimes do differ from one another, three different sets of measurements of sea level rise have reached the same conclusion: since 1900 it has been about 7.9 in., while the rise from 1993 on is about 4.5 inches, which is an increase of 3.57 mm/year since 1993.[2]

At first glance, these numbers may seem innocuous and not of consequence because the increases seem to be small. The problem of drawing such a conclusion is that, just like global temperature averages, the averages of sea level rise say nothing about the deviations from them, which include the washing away of land and the higher tides that can and do lead to episodes of flooding, often with devastating impacts.

I have selected a representative group from around the world to illustrate this phenomenon, starting with three from the USA and ending with the west coast of Africa, the worst of all these situations.

Along most of the East Coast of the USA, sea level rise relative to its 1950 value has increased the height of the tides by between 5 and 15 inches. It should therefore not be surprising that there are instances where the occurrence of this rise has led to some very serious consequences.

One example is the 2012 tropical storm Sandy. Its storm surge over lower Manhattan and Staten Island was about 9 feet, of which 1 foot was sea level rise. About $19 billion worth of damage occurred; the water covered an area of over roughly 51 square miles and inundated lower Manhattan subway stations and tunnels.

Miami Beach, where sea level has risen about 1 foot since 1930, is unfortunately built on porous limestone through which sea water enters the city. This increases the flooding caused by so-called king tides. The flooding has impacted

both septic tanks and sewer systems in Miami-Dade County, with effluents that have turned some of the flood waters toxic (e.g., laden with virus and human excrement). In an ironic twist, while the flooding has been higher in some poor African American neighborhoods, which are the least likely to get relief, some of those areas have caught the attention of real estate developers who apparently have plans to build high-rise buildings there. Attempts to combat the flooding by raising streets, installing pumps, and building retaining walls are estimated to have cost as much as $500 million. While much more is needed for the future, it is not yet known whether these efforts to stem the slow increase of flooding due to rising sea levels will be sufficient.

An analogous situation exists in Norfolk, Virginia, which has had its share of flooding as well as sinking land. Flooding of streets is exacerbated by sea levels rising, where the rise has been about 6 inches since 1992, which is roughly twice the global average. The city has raised streets and installed pumps and floodwalls, but it will be difficult to do this in the poorer regions of the city, and it will quite possibly not suffice in the long run.

The situation is worse at the US Naval Station Norfolk, which is adjacent to Norfolk. Sea level rise there has been about 15 inches in the last 100 years, and that, coupled with both storm surges and slowly sinking land, has led to unavoidable consequences, despite the efforts that have been and are being made to mitigate them. For example, the Naval Station is flooding about 10 times per year due to high tides made worse by sea level rise; this flooding makes the roads to the base impassable, which will become even worse over time as sea levels continue to rise.[3]

Sea level problems are not limited to the USA: sea level rise is threatening both large and small cities bordering on or in the Pacific, as well as numerous low-lying islands there.

Indonesia announced in 2019 that because the land under its capital Jakarta is sinking and sea levels are rising (about 40% of the city is under sea level), it would relocate its capital to the island of Borneo. The planned relocation has already raised environmental concerns, ones that seem likely to be ignored. In Kiribati, rising seas are already washing away the edges of the atoll where most of its population lives and is contaminating the aquifer that supplies its drinking water. Plans exist to relocate its entire population, assuming all can be accepted much less accommodated. An equivalent situation exists for the Maldives archipelago in the Indian Ocean, where sea level rise, already affecting the archipelago's islands, will eventually submerge them and the US base in Diego Garcia. The native population will then join millions of other climate refugees.

Rising seas will wash away the land in the Orkney Islands where recently discovered archeological sites are located. On Easter Island, whose statues are an archeological gem, rising seas have produced waves that are encroaching on this site.

The Italian city of Venice, which also is an archeological treasure, has experienced its own share of flooding from rising sea levels. A severe instance occurred in October 2018: very high winds increased the high tide about 5 feet above the average level; the water also rose up through the paving stones of the streets and piazzas. It was the biggest flood in 10 years, with water entering many shops and restaurants as well as historic churches, especially St. Marks Basilica, which had already suffered significant water damage high up on its walls. And then, a little more than a year later in November 2019, even greater flooding occurred. Such flooding will worsen because the city's buildings are very slowly sinking (as they have been for centuries, but now it seems impossible to counter this by building new ones on top of them, as had been previously done).

To defend the city and its treasures, it was decided to install flood gates at the bottom of Venice's lagoon, gates that could be raised when flooding was predicted to be imminent. After years of delays, the project was sufficiently complete that it was employed for the first time against a 4-foot tide on October 3, 2020. The test was successful: 78 floodgates were raised; the city, especially St. Mark's piazza, remained dry. Next day, however, a tide of about 3.5 feet flooded the piazza to a height of about a foot, while currently it floods about 60 times per year, with a notable flooding occurring in November 2021. Unfortunately, the gates do not always function properly, while they will only be raised for an expected tide of at least 4 feet. It was anticipated that the gates will be up for possibly half the year, which could cause the organic matter in Venice's lagoon to die. There is also uncertainty if the gates can withstand very high winds and whether they can consistently function as needed. Whether similar measures will work elsewhere is unknown.

Parts of the mainland of Asia have also experienced the dire repercussions of sea level rise. An extreme example is Bangladesh[4], whose southernmost portion ends at the Bay of Bengal. It is a delta formed by the convergence of three rivers (the Brahmaputra, the Ganges, and the Meghna) and is the home of one of the largest mangrove forests in the world. Since sea level rise there is one of the most rapid anywhere in the world, it has been partly responsible for huge losses of the mangroves and has helped storm surges from cyclones to propel water up the rivers as much as 50 miles or more. Because these events have contaminated agricultural land and ground water, there have been migrations to India, which does not want these refugees. How many of the delta dwellers have attempted to migrate is unclear, but if they are Muslims, the anti-Muslim law in India (as it is perceived) will likely prevent this. Such migrant attempters will be added to the tens of thousands that have had to be temporality relocated into poorer living conditions during such events. Attempts to counter the flooding by constructing walls have been successful to some extent so far, but the future is uncertain, since much of this part of Bangladesh is only a little above sea level. It is the opposite of the Netherlands, much of which is below sea level and has been successfully protected by dikes for centuries, though there are concerns that these are on the way to becoming inadequate.

The flooding that has irrigated the streets and piazzas of Venice occur from a combination of sinking land and rising sea levels that increase both king tides and the height of storm-driven water entering its lagoon. The city's area is about 415 km^2 (160 square miles); its population is estimated to be 270,000. For many persons in the USA and Europe, Venice is an archeological treasure that is irreplaceable. It is likely that most of these people would regard its flooding as a tragedy, so it is not unreasonable to wonder how they would regard the devastation of the flat, 6500 km (4000 mile) long coastline of western Africa from Mauritania to Cameroon, where roughly 30% of their populations live. As summarized by Goodell,[4] the tens of millions who live in the coastal cities, towns, villages, and isolated dwellings, a large majority of whom are poor, are being profoundly affected due to contamination of drinking water and agricultural soils, rainy season flooding made worse by sea level rise, overrun sewage systems (where they exist), erosion of land (in some places as much as 100 feet in a year), and untreated diseases. If Venice represents an archeological tragedy, this is much worse, a human tragedy occurring on a scale larger than is the case so far in the Pacific, much less Venice itself, but what is likely to occur in both developed and developing countries.

Ocean Effects

Sea level rise is not the only effect global warming has had on the oceans. Another is on coral reefs. Not only does approximately 90% of the heat generated by CO_2 increases enter the oceans, but a recent study found that between 1991 and 2016, the average amount of this heat per year is roughly 60% greater than previously estimated. The resulting rise in sea water temperatures will therefore imperil coral reefs more than had been thought. The additional CO_2 brought into the oceans makes the sea water more acidic. This has a double effect on corals: loss of their algae (that live symbiotically with the coral) and degradation of the coral's pigment. If the algae are not replaced or the pigment is not restored, the coral will die.

Why is this important? Because coral reefs are the home of roughly 25% of marine life (more than a million species), some of which are predators of other reef inhabitants, while others are on the chain of marine life that leads to the fish consumed by people. Dead reefs are no longer the home of living creatures; reduced-in-size coral reefs that do not recover—and these are predominantly ones in shallower waters—will eventually mean a reduction in the amount of fish available for human consumption. The reduction or death of corals has been observed in different parts of the world, for example, Australia's Great Barrier Reef, while nearly 50% of Florida's reefs have disappeared, these reefs being the third largest in the world. Although there are other factors that negatively affect coral reefs, global warming is one that humans can, at least in principle, try to mitigate.

Global warming also impacts hurricanes. During summertime in the northern hemisphere, GW heat can raise ocean surface temperature by 3 °C or more above the average, which in recent years has affected hurricanes in two important ways. First, the surface temperature increase causes more water than normal to evaporate, adding to the amount in hurricanes and already in the atmosphere. This has led to higher hurricane wind speeds and thus more category 4 or 5 storms. Second, increased surface temperatures typically enlarge a hurricane so that it covers more area over the ocean, and thus over land when it goes ashore. The combination of increased extent and additional water means that if it does go ashore, there will be greater inundation of the land. This will occur even if the hurricane is downgraded to a tropical storm. The devastation caused by this increase in its water content (as well as its winds) can be and has been disastrous. A few recent examples of such inundation in the USA, in some cases with toxic consequences, are parts of Florida, of North and South Carolina, of Puerto Rico, and the cities of New York and Houston. In the Pacific Ocean, some cyclones have been equally deadly, such as those that have hit parts of Japan and China, and especially the Philippines (e.g., Typhoon Rai in 2021).

But such deadly storms are not confined to the Atlantic and Pacific oceans. They have also struck portions of India and Bangladesh that border on the Bay of Bengal, where, as noted above, sea level rise has had devastating effects. That misery was added to in May 2020, when super cyclone Amphan hit, though it weakened somewhat when it came ashore. With millions of people previously evacuated, it devastated infrastructure, destroying newly planted crops and tens of thousands of dwellings in both countries, as well as damaging millions of houses. Its storm surge inundated fresh water sources with salt water; the livelihoods of over a million people were adversely affected, with poverty-stricken people, especially but not exclusively in Bangladesh, made even more so. Trees and bridges were swept away, and livestock killed. It continued inland, with its storm surge and heavy rain flooding streets in Kolkata (formerly Calcutta), destroying buildings and trees, and leaving 14 million residents without electrical power.

While Bangladesh and India are on the northern portion of the Indian ocean, in January 2022 tropical cyclone Ana arrived on its southern portion, striking the southern African countries of Madagascar, Malawi, and Mozambique. Over 450,000 people were displaced and made homeless, with infrastructure such as roads and medical facilities badly impacted.

Like the coronavirus pandemic, the repercussions of global warming are obviously felt worldwide, as the following sections will continue to illustrate. What needs to be kept in mind is that developing and developed countries can—at least so far—deal better with the consequences of global warming than can less well-off ones, of which Bangladesh is almost a poster child, one we shall encounter in this regard again.

Glacier, Sea Ice, and Ice Sheet Repercussions

In addition to global warming having affected the oceans, it has impacted sea ice, ice sheets, and glaciers. Almost all glaciers on land have noticeably retreated (some completely) and are continuing to do so. An example is Glacier National Park in the USA, which had about 150 in 1850 but was down to 25 in 2020. While this is another indication that global warming is occurring, rejecters have argued against glaciers retreating worldwide without, of course, presenting any supporting evidence, since none exists.

The major consequence of the loss of glacier ice and the concomitant decrease in the amounts of snow and of ice sheets has been and will continue to be a slow diminishing of the water they provide for agriculture, for some instances of hydropower, and a significant, even disastrous reduction of fresh drinking water. The foremost example of this is the Tibetan Plateau, which, as noted earlier, is heating up to three times faster than the global average and is the source of 10 of the earth's biggest rivers, whose waters sustain over 1.5 billion people[5] (more on this in the next chapter).[6]

Many less dramatic but no less important examples exist, one being the loss of a glacier in the Andes that has adversely affected a much smaller and indigenous population. The decrease in water for agricultural purposes will lead to certain foods no longer being grown, to food shortages, and to wars being fought over it. The shrinking of glaciers has already forced successful and profitable farming in some regions to return to growing less profitable crops. In those regions, the end result will be a complete loss of livelihoods and involuntary relocation to areas where such refugees may well have great difficulty being able to find satisfactory work—if any at all.

Loss of both sea ice and ice sheets is another consequence global warming. Sea ice loss is mostly confined to the Arctic, which, as noted previously, is warming three to four times as fast as the rest of the globe, as vividly exemplified by the June 20, 2020, record temperature value of 38 °C (100.8 °F) at the small Siberian town of Verkhoyansk (inside the Arctic circle), whose average June temperature has been 20 °C.

The area of sea ice decreases as the Arctic warms up, with the least amount occurring in September. Different measurements give slightly different results, but the typical summertime minimum area has declined about 40%

in the last 40 years, with the largest decrease occurring in 2020.[7] In addition, it has become possible to estimate the decrease in the volume of Arctic ice as well, which is a staggering 73% over approximately the same period.[8]

Since sea ice floats, its decrease will not raise sea levels. In fact, it might be thought to be a positive development in the near future: opening a summertime route and eventually a year-round one through the Northwest Passage in the Arctic to the Pacific and the Northeast Passage above Russia, as well as opening of the Arctic to the mining of valuable mineral deposits. The first would be profitable to shipping companies; the latter has already led to conflicting claims between countries as to who has the rights to these minerals. However, there are major down sides to the loss of Arctic sea ice that may not come readily to mind, ones that are examined below and in the next chapter.

In contrast to sea ice, ice sheets are on land, are very extensive, and include mountains and valleys of ice. The ice sheet of Greenland covers an area of about 1.7 million km^2 (about 3 times the area of Texas) and if all of it melted sea levels would rise by about 7 meters. Antarctica's ice sheet on the other hand covers an area of roughly 14 million km^2 (about the size of the USA and Mexico together) and would raise sea levels about 61 meters if all of it melted. These two sheets are estimated to contain about 99% of the planet's freshwater. In contrast to ice sheets, ice shelves are in the ocean, sited on the land underneath them, and usually are connected to ice sheets, mostly in Antarctica. They act as barriers to the sheet's losing ice through some of it sliding into the ocean, as has already occurred.

Measurements from 2002 forward show that the ice sheets of Greenland and Antarctica have each shrunk. The amount for Greenland is roughly 5200 billion tonnes, and while there have been many relatively small up and down fluctuations, the trend is unmistakably a decreasing one, with the greatest loss being over 530 tonnes in 2019, with slightly less in 2020. This is a direct consequence of GW (see comments below), which has had different effects in Greenland that are nicely illustrative. One of them was the unprecedented occurrence of rain falling over Greenland's highest ice peak in August 2021, when the temperature was about 18 °C above the average! Another is the melting of the ice sheet, which in July of 2021 totaled nearly 17 billion tonnes over three days, emblematic of greatly increased melting that has been occurring there. Note, by the way, that 17 billion tonnes of melted ice would cover the entire US state of Florida by roughly two inches or 5 cm of water.

In contrast to Greenland, Antarctica's overall ice loss is over 2800 billion tonnes. Recent research has ominously found an acceleration to this shrinkage, with the total loss so far increasing sea level by about 0.6 inches. For example, its Thwaites glacier has been losing about 45 billion tonnes per year in recent years. It is the one that scientists are most concerned about, though when it could significantly collapse, causing greatly increased sea level rise, is unknown.

While the overall decrease of Antarctica's ice over time is also unmistakable, its fluctuations have been much more dramatic than Greenland's smaller ones, which are reminiscent of those on the Keeling curves. Antarctica's are more dramatic because some of them have temporarily and strikingly increased, after which they have declined to a lower value. An example is the big spike that occurred in 2016, which was then erased by the amount of ice lost in 2017, thereby continuing the overall decrease. Antarctica's fluctuations are so different than Greenland's because Antarctica usually receives much greater amounts of snow, some of which becomes ice.

Given this background, does GW play a role in these ice sheet losses? The answer is yes. The loss in each of them has recently been accelerating, with that of Greenland being the more rapid. West Antarctica and the Antarctica Peninsula have been losing ice for some time, whereas East Antarctica has been gaining ice, but not enough to offset the losses. Global warming promotes the losses and the accelerations in two ways. First, it warms the air above the ice sheets, which in turn warms the surface enough to cause melting. Second, the warmed seas next to the edges of Greenland's and Antarctica's ice sheets have warmed them enough to have caused icebergs to break away. In addition, the warmed seas next to some of Antarctica's ice shelves have caused them to detach, exposing the sheet itself to these seas, thus freeing portions of the sheets to begin a slow sliding into the water. These effects are almost certainly likely to continue as more CO_2 enters the atmosphere. On top of this, as if to compete with the Arctic temperature increase, since the 1990s surface air temperatures at the South Pole have been warming about three times greater than the global average, with temps in some places there 40 °C above normal in March 2022. These are, of course, increases to which anthropogenic global warming has contributed.

Land Impacts

Global warming impacts on the earth's land are consequential both locally and globally. Included are heat waves and extremely high temperatures; migration of different species northward in the northern hemisphere and their extinction in whole or in part; longer growing seasons; more intense forest fires; torrential rainfalls and flooding; extended areas of drought plus the creation of new ones; and shrinking glaciers.

While glacier retreat has already been examined and identified as anthropogenic, it's included here because of two instances involving it that rejecters pounced on. The first concerns the false claim that Al Gore in "An Inconvenient Truth" asserted that global warming had caused the retreat of the glacier on Mt. Kilimanjaro, whereas it had been losing ice well before Gore's supposed assertion. In fact, he did not assert that GW had caused the retreat, acknowledged that it had lost much ice before 1970 and referred to a quote by an expert on mountain glaciers that the ice on Kilimanjaro would be totally gone within 10 years (which, as it turns out, didn't happen). The other false claim involved an error concerning glaciers in an IPCC Report. It was the *only* identified error in it, but the rejecters claimed that the error invalidated the entire report! As one climate scientist has noted, his and other climatologist's work and reports on GW must be right 100% of the time, whereas a mistake of 0.01% becomes invalidating fodder for rejecters.

The role of global warming in most heat waves has been an indirect but substantial one. There are different definitions of a heat wave, all of which note that it is a prolonged time period where the temperatures are noticeably higher than normal for the region experiencing it. The different definitions concern how long the prolonged period lasts, ranging from at least two days to one week or several, and recently for longer. Most climatologists accept that their cause is a stationary high-pressure system over the affected region, resulting from the jet stream being slowed and then stalling. The jet stream is the current of air that flows from west to east in the mid-latitudes of the northern hemisphere. It can dip southward at times. The normal jet stream has a lower temperature on its northern side and a higher one on its southern side, with the former resulting from colder arctic air flowing south and the latter resulting from warmer tropical air flowing northward. Where they meet defines the stream. Its slowing occurs when the cold air flowing south is warmer than usual. The north–south temperature difference then becomes smaller, which causes the slowing and eventual stalling that leads to the high-pressure system noted above. This high pressure not only increases the warm air below it but also keeps it in place, leading to the heat wave, which can last a much longer time than just a few days (and often does).

The temperature increase on the north side of the jet stream is an indirect effect of GW. The direct effect is the higher temperatures of the Arctic, which is warming much more quickly than most of the globe and melting its sea ice, as noted above. That melting decreases the albedo from the Arctic, increasing the amount of solar insolation striking it, thereby adding to the warming. The overall result is the temperature increase on the northern side of the jet stream, so that GW insidiously helps to create the heat wave.

Thousands of people have died from prolonged heat waves, making it much more than a "normal" situation of increased temperatures. Some of the very high temperatures it has produced recently are 50 °C (122 °F) and greater, as in western Australia in January 2022, and in India and Pakistan in May 2022. Such temperatures melt road surfaces and cause the blistering of pets' paws if they are in contact with roads or pavements. More importantly, at these high temperatures human cells begin to cook, blood thickens, muscles lock around the lungs, and the brain suffers from a decreased amount of oxygen. These effects are lessened by human perspiration but if the air itself is humid, this bodily defense weakens.

Heat waves are clearly far more serious than simply an added discomfort (or the cost of air conditioning). An estimated 70,000 deaths occurred in Europe in the summer of 2003 and 56,000 deaths in Russia in the summer of 2010. Other major ones have happened since. Climate researchers have concluded that human influence (AGW) more than doubled the risk of the 2003 heat wave occurring, while the probability is 80% that the 2010 one would not have occurred without GW.

Similarly, the exceptional summer heat wave of 2017 in southern Europe was determined to be 10 times more likely than would have been the case in 1900, while the size of the one in the summer of 2018, which spread from Japan to Canada, was unprecedented and would not have been possible without global warming. The European heat wave in the summer of 2019 that produced the highest recorded temperatures in the UK, Germany, Belgium, and the Netherlands is thought to be 100 times more likely than if global warming had not occurred. Even more troubling is the January to June 2020 heat wave in Siberia, whose chances of occurring were estimated by climatologists to have increased by roughly a factor of 600 because of global warming. Average temperatures were pushed about 5 °C higher than normal, with the 38 °C record being set then in the town of Verhoyansk. This prolonged heat wave led to massive wildfires in June, which released an estimated 51 million tonnes of CO_2, more than the *annual* emissions of smaller industrialized countries such as Switzerland.

Heat waves are continuing in many places globally: record temperatures occurred in various places in Europe in June 2021; it was 67 °F = 16 °C in Kodiak, Alaska in December 2021; while in much of northern China in June 2022 it was 40 °C, with asphalt road temperatures in Henan Province reaching 165 °F = 74 °C! This was followed in July 2022 by a record setting heat wave in China, in much of western Europe and in the USA, with temperatures so high in places (mid 40s °C) that railroad tracks buckled, causing some train services to be put on hold. Accompanying this were intense wildfires, which, coupled with the high temperatures, led to thousands of deaths in a very short period of time. There is every expectation that this is only the beginning, as discussed in Chapter 3.

An analog of the typical heat wave is a new phenomenon known as a *heat dome*, which occurred in June 2021. It is thought to have been caused by a GW tropical storm that changed the jet stream over British Columbia (BC) in Canada and the US states of Oregon and Washington. The change was a broad groove or dimple in the form of the capital Greek letter omega (Ω) that stayed in place for several days, not only increasing and trapping the heat below it but preventing cooler air from the Arctic to enter the dome's region. The result was a vast increase in air temperatures in these areas, hitting highs close to or even higher than 50 °C. Forest fires broke out in portions of BC, where the heat-related death toll was 486, while at least 95 deaths were reported in Oregon and in Washington, though this number is likely an underestimate. While these high air temperatures help explain the deaths, they are not the only measure of the impact of the heat dome. As reported in a New Yorker magazine article, an infrared camera measured the temperature of asphalt in Portland, Oregon, and found it to be 82 °C = 180 °F! And just as in northern China in June 2022, asphalt roads buckled, and people who were near any of it suffered badly.

Heat waves are one of the impacts of global warming on the earth's land surface. Another is the movement in the northern hemisphere of animals and insects, including birds and butterflies, into areas farther north than their traditional habitats. However, certain species are trapped where they live and cannot migrate. To survive, some plant populations likewise must move north—if they can. Sooner or later, if some species cannot adapt, they will become extinct. Extinction of some species has already occurred in local habitats, but has apparently not occurred globally, though if temperatures continue rising, as expected, this is very likely to change, with global extinctions becoming a reality.

A pestiferous consequence is the movement northward of bark beetles, which eat the bark of trees, eventually killing them. Warming has meant that the beetles' larvae are not dying off as quickly as previously, that more of them hatch, and that their killing power has increased. These beetles have attacked both pine and spruce trees in different parts of the world, e.g., western USA, with many thousands of acres affected. The dead trees then act as fuel for forest fires, recently made worse because of higher temperatures.

A further influence on forest fires is the drying out of brush, fallen leaves, etc., which adds even more fuel. Under normal circumstances, forest fires are a means by which new forest growth can eventually occur, but these are deadly not normal times. Examples in which global warming is a contributing factor are the giant ones in Siberia, in California and other western US states and in Canada that have destroyed whole communities, and the bush fires in Australia that began in 2019 and ended in February 2020. Australia's demolished roughly 85,000 km^2 of forest, killed an estimated 1 billion animals, and destroyed nearly 6000 buildings. These numbers are so large they almost (but do not!) make the 34 human deaths seem inconsequential. It was the most devastating wildfire in modern Australia's history. The acreage being destroyed in California and Siberia is also gigantic. This situation is highly likely to get worse, given that anthropogenic CO_2 is increasing, as exemplified by the horrific Dixie fire in California in 2021 and the ones that are continuing there.

Although this chapter is on repercussions of global warming, with wildfires being one of them, there is a type of converse effect on GW from them and from deforestation: these events not only add CO_2 to the atmosphere, but also they also reduce the number of leaves and plants that can reabsorb it. This can impact the seasonal fluctuations seen in the Keeling curve, such that the absence of these latter absorbers means more CO_2 remains in the atmosphere; otherwise, it would have been removed. In the long run, this will lead to an increase in T_{av}, though not a large one. One place where wildfires have been set and deforestation has intentionally occurred is the Amazon rainforest. If the Keeling curve fluctuations are an analog of breathing, the Amazon rainforest has conservatively been estimated to be about 8% of the earth's lungs, so these deliberate activities are equivalent to decreasing the earth's lung capacity, although only by a small amount. Enough deforestation has occurred in the Amazon rain forest that as of July 2021, it was emitting more CO_2 than it was absorbing. Such deforestation has occurred in many countries and not just the Amazon, being done deliberately for various reasons, almost all commercial. In the long run—though how long the long run may be is uncertain—continued deforestation will lead to increased global warming. To counteract it requires reforestation, considered in the next chapter.

Another impact of GW on the earth's land surface is the drying out of soil. The resulting evaporation adds moisture to the atmosphere, leading to increased rainfall, which almost always does not fall on the parched land, leaving the parched soil to become even drier and eventually a desert. The increased rainfall can occasionally be torrential, leading to what can be massive flooding. Such drought and flooding have occurred in different parts of the world, including a multiyear drought in roughly 45% of the lower 48 states in the USA, resulting in the driest two decades in the past 1200 years in the US southwest. Africa has experienced drought for some years, wherein agriculture has been perilously affected; in mid-2020, roughly 45 million people from over ten countries in southern Africa were food insecure. Several million had previously been added to the world's hunger refugees. That should not come as a surprise, since so much of the population there lives in poverty, with whatever help has been provided proving inadequate. In addition, past droughts have affected the northern portion of Africa killing over 100,000 people and distressing well over 50% of the farmland.

The global warming pattern of at least a several-year drought followed by heavy rains wiping out recently planted crops followed by another drought as seen in Africa has also occurred elsewhere, for example, in central America, with Guatemala being a current example, where it has impacted food supplies and leading to a not-unexpected migration northward.

A different situation, analogous to drought, occurs when the rivers that feed lakes and inland seas are either diverted or diminished, whereupon the lakes and seas begin to vanish. Among numerous examples are the Aral Sea, Lake Urmia, and Lake Chad. Global warming enters these situations by inducing changes in rainfall patterns which lead to little or no rain being available to resupply the vanishing waters. A different scenario has caused Lakes Powell and Mead in the USA to sink to their lowest known levels, which means that their ability to supply electricity from the dams that form them is at major risk. These lakes are fed by the Colorado River, which is itself at extremely low levels due to lack of both winter snowpack and rain in the mountains where the river originates, lacks due to global warming. This not only has impacted the lakes but is a huge problem itself because 40 million people rely on the river for various agriculture uses in seven different states as well as on American Indian tribal lands and even in northern Mexico. Continued diminution of the Colorado River will very likely mean that food production from these areas will begin to decrease, an effect that will be felt in other parts of the country, since California especially is a major supplier of fruits, vegetables, and wines. Given that the Colorado River is already so low, the big west coast snowfall of late 2021 has not been able to alleviate this problem. Because of water rights that were established long ago, this will likely be a difficult problem to resolve between the affected areas.

This is not a problem limited to western North America, two other examples being the high heat and low winter snowpack that has caused the Rhine River in western Europe to become perilously low, while the Parana River in South America has sunk to drastic lows due to GW-induced drought.

While global warming-related droughts and torrential rains can occur in tandem, droughts need not accompany such rains, as in the case of the inundations that occur when hurricanes that have absorbed more than the usual amount of water make landfall, as discussed above. Very recent studies indicate that global warming can similarly influence monsoons, for instance, those in Asia that have been the worst in many years. Monsoons and their heavy rainfall are a yearly phenomenon, but these latter ones were especially severe, leading to extensive flooding in China, Bangladesh, and portions of Nepal and northeastern India, starting in June 2020. They are continuing to occur. Over 550 people have been killed in these latter three countries, while well over 9.5 million have been directly affected. At least a third of Bangladesh was estimated by its Ministry for Natural Disasters to have been underwater, which has occurred in a country previously ravaged by super cyclone Amphan. With homes destroyed, plants and animals again liquidated, recovery will be slow if not problematic. Bangladesh clearly exemplifies how countries that are among the least responsible for global warming tend to end up bearing the unwarranted brunt of its repercussions.

China was even more badly hit by flooding from the monsoon's torrential rains. As of early August 2020, the Chinese Ministry of Emergency Management had estimated that nearly 55 million people from 27 provinces had suffered from what has turned out to be record breaking floods, the worst in over 20 years. More than 40,000 houses were devastated, 13 million acres of cultivated land were impacted, and nearly 158 people were dead or missing. An even worse situation occurred two years later: as of late summer 2022, a monsoon and glacier melt flooded one third of Pakistan, killed at least 1700 people, destroyed hundreds of thousands of buildings including homes and schools, devastated agriculture and food supplies, thereby disastrously affecting 33 million persons. The worst disaster in the country's history, it is another event to which global warming has contributed.

Furthermore, heavy flooding needs not be the result of monsoons: GW-induced torrential rains can cause it, as has occurred in Australia and various European countries very recently, some instances being England, Germany, Luxemburg, Belgium, and Switzerland, as well as in Bangladesh and India, while the flooding in southern China in June 2022 affected "only" 479,600 people, ruined crops, and collapsed at least 1700 houses.

The foregoing recounts some of the repercussions from global warming, which is mainly caused by the burning of fossil fuels. In addition to emitting gases such as CO_2, this burning also emits aerosols, which are liquids or small solids suspended in the air. Those containing infectious agents and hazardous materials are dangerous to humans if they are inhaled, as they can then penetrate the respiratory tract and unfortunately remain there for too long a time. The effect of these toxic elements on humans has serious consequences and has already occurred in various parts of the world, especially in Beijing. Moreover, the burning of fossil fuels is not the only anthropogenic source of aerosols. Others include the exhaust from automobiles that had severely impacted New Delhi, the exhaust from aircraft and from biomass burning in developing and poor countries. The repercussions are obviously global and can be severe; however, COVID-19 has changed this significantly, e.g., the air in many cities has become much cleaner, as have Venice's canals. How long this will last is uncertain. See Chapter 3 for further comments regarding the COVID-19 pandemic's consequences and the future.

Precursor Repercussions: Fossil Fuel Extraction and Transportation

It is not only the burning of coal, oil, and gas that has led to deleterious consequences: their extraction and transportation have done so as well. This remains true for coal, despite the significant decline of mining in the USA since it is a major industry elsewhere, used in those parts of the world that continue to rely on it to generate electricity.

Underground coal mining has long been one of the most dangerous kinds of work: the toxic gases released during mining cause black lung disease, a horrific respiratory ailment, while explosions from coal dust have collapsed mines. Death is not an unusual outcome. Above-ground problems also occur: the toxic gases and coal dust can seep into the surrounding atmosphere, infecting nearby residents, while the collapses have led to the subsidence of the land overhead and in its immediate vicinity, injuring people and destroying buildings.

While these are known consequences of an activity in the USA in which the miners are well paid and the owners have amassed huge profits, there are, unfortunately, deleterious repercussions that are far more widespread, some resulting from above-ground strip and mountain-top mining.

A different environmental impact of both under- and above-ground coal extraction and of the burning of coal in power plants has been turning streams and rivers toxic, killing fish and plants, and impacting drinking water. It is caused by coal interacting with air and water, producing sulfuric acid, which, in combination with metals such as copper and mercury, seeps into nearby waters, turning them toxic.

In addition to the emission of greenhouse gases when coal is burned, its extraction releases the greenhouse gas methane (CH_4); roughly 10% of the methane produced in the USA has been emitted this way. It is a small contribution to the nearly 40 million tonnes of CH_4 that were released from new and disused coal mines globally in 2018, as determined by the International Energy Agency (IEA). That tonnage is comparable to the methane emissions from aviation and shipping combined. As a more potent greenhouse gas than CO_2, its effect is dependent on the time period considered, as discussed in the next chapter. The IEA estimated in its report that the global warming impact of 1 tonne of CH_4 is equivalent to that from roughly 30 tonnes of CO_2. This is not an insignificant number.

While coal extraction is land-based, oil is obtained from drilling on land and under water. Each type has led to environmental disasters, with oil spills from offshore oil rig failures having received the greater press and media attention. Although oil rig catastrophes have occurred in various parts of the world, e.g., off the coasts of the UK, Norway, China, Canada, Thailand, Brazil, and elsewhere, the Deepwater Horizon spill in the Gulf of Mexico in 2010 is the one most Americans would probably think of if asked to specify an example of this type of "problem." It is not only the most recent of several in the Gulf of Mexico, but it is the largest oil rig spill in history. Although it took five years to cap the well, that did not stop the leaking, which has produced an amount of oil estimated to be at least 650,000 tonnes that has killed fish, maimed and killed aquatic mammals, and contaminated beaches. Unfortunately, it is not the longest that has occurred, an "honor" that belongs to the 2004 Taylor oil spill in the Gulf of Mexico, which was almost completely capped in 2019: the small amount that was still leaking—about 1000 gallons per day, as of November 2021—was being brought to the surface and captured. An upper-limit estimate of the amount leaked before partial capping is 460,000 tonnes, not as bad as Deepwater Horizon, but still ...

Oil obtained from offshore rigs and from drilling in countries with no refineries is transported to refineries by oil tankers. It should be no surprise that oil tanker groundings, collisions, and shipwrecks have also caused oil spills that damage coastlines, kill fish, etc. They have occurred in numerous places around the world, of which the Torrey Canyon shipwreck in 1967, the Exxon Valdez grounding in 1989, and the Prestige sinking in 2002 are prime examples. Since the tankers carry very large volumes of oil, the sizes of the spills can be considerable, with those from the Prestige and the Exxon Valdez being 60,000 and 37,000 tonnes, respectively.

Although drilling for oil on land began in 1859, it took 90 years for the newest method to be discovered and employed. Known as hydraulic fracturing, and commonly referred to as "fracking," it is used to extract both oil and natural gas. Doing so has led to dire repercussions, which is the final topic in this chapter.

The first step in hydraulic fracturing is to create a cylindrical opening (a "borehole") by drilling deep—many thousands of feet—under the earth's surface to or near a region where so-called shale gas or oil is thought to be located. The drilling goes through rocks and soil, presumably below the local water table. A million gallons or more of liquid under very high pressure is then forced into the borehole to fracture the rocks, creating cracks and openings through which gas or oil can flow and be extracted. In some situations, after creating the borehole, the drilling apparatus is turned sideways and the drilling continues horizontally for as much as a mile or more to a new site where gas or oil is expected (or hoped) to be. The liquid-induced fracturing occurs along the newly created, horizontal borehole, again allowing the gas or oil to flow up and be extracted.

Along with its huge volume of water, the "slickwater" fracking liquid contains gels, chemicals, and enough sand (injected at the proper time) so that the combination of sand and gels will keep the cracks open. Unfortunately, the injected liquid does not stay in place: it joins the gas or oil and upwells to the surface; because it is toxic, it must be disposed of safely, but that has not always been the case, as described below.

Natural gas, which consists mostly of methane plus smaller amounts of other molecules, is less expensive to extract and transport than is coal, which has been the main reason for the decline of coal mining in some parts of the world. Natural gas has been touted as a "bridge" to a future in which burning it reduces the amount of greenhouse gases emitted as compared with the CO_2 emissions from the burning of coal and oil. It has thus been claimed as having a noticeable, positive impact on climate change.

However, there is a major problem with this claim. A recent study reported that in addition to the methane produced when the natural gas is burned, the leaking of natural gas from some of the wells where the fracking takes place has caused significant "climate pollution"—i.e., methane escaping into the atmosphere: in one case, 11 times the equivalent CO_2 emitted by coal fired plants, and in another approximately 12 times as much. The former number was revealed to be much larger than the amount that the oil and gas companies had reported to the state where the fracking had occurred (Pennsylvania). It is disquieting that this phenomenon had been identified in scientific studies published as early as 2011 yet has not received much attention outside of scientific circles, since doing so would presumably counter the claim of natural gas being so much more climate friendly. Until this type of CH_4 emission is capped and/or captured, natural gas will not quite be the "bridge" it has been touted as.

And this is not just a problem of escaping gases: there is also the phenomenon known as *flaring*. Flaring is the deliberate burning of the extra natural gas that remains when more of it is extracted than can be carried away by the pipelines that move it to energy-generating facilities. In addition to being wasteful, it can add significantly to the amount of methane entering the atmosphere. It is an especially egregious problem because gas-producing companies have known about it, and while they "should" be capturing it, increasing amounts have been burned.

These methane releases and burnings are not the only negative repercussions of fracking. Another is contamination of aquifers by the re-injection, supposedly below the water table, of the upwelled toxic slickwater, as well as leaks of it from trucks intended to transport it to safe places, and from tanks where it was considered safe to store it. Such leaks have been ingested by cattle, killing them, as well as infecting people who have inhaled their toxic fumes.

As if this is not enough, another consequence of fracking is triggering earthquakes. Earthquakes are normally a natural phenomenon, mostly caused by one of the earth's tectonic plates trying to move past another; they occur worldwide, with one estimate being half a million per year. Being natural, they are unavoidable. In contrast, evidence indicates that fracking-induced ones are not unavoidable: they are, in effect, anthropogenic. Such earthquakes occur in two ways: via drilling and via re-injection of the slickwater. According to the US Geologic Survey (USGS), re-injection is the major cause. They have occurred in various states in the USA, especially Oklahoma, where a survey of them indicated a 10,000% jump in the number that people had previously experienced. This Oklahoma earthquake "swarm" began in 2010: over the 30 years before then, there had been about 40 that were greater than magnitude 3.0 on the Richter scale. Fracking changed that, with more than 20 magnitude 3.0 earthquakes occurring in the first half of 2010. As Oklahoma accommodated the cash-cow drillers by not imposing restrictions on hydraulic fracturing, these 3.0 numbers kept increasing, reaching a high of 900 in 2015. Steps were finally taken by the state to control fracking, with the result that in 2018 there were "only" 166 of the 3.0 quakes but declined to 29 in 2021. Earthquakes have continued there, averaging about 5 per day from January 2021 to January 2022, with the biggest one since 2016 occurring in January 2022: magnitude 4.5. Overall, thousands of fracking-induced earthquakes have occurred in Oklahoma and adjacent portions of Kansas and Texas.[9] They have, of course, happened elsewhere in the USA.

While the costs of starting a fracking operation are very large, the initial profits to be made are as well, so it has by no means been confined to the USA: its negative repercussions have been experienced in more than a dozen other countries. Some have already banned it, for instance, France and Tunisia, while the Netherlands is to do so in 2022, and the UK has placed a temporary moratorium on it at the only site where it has been carried out. A reason for not yet banning it elsewhere is the amount of money that countries/states get from the drillers; in some of these places, little or none of this income is spent to improve the lives of its citizens: the rulers/dictators keep it for themselves and their families.

This outlining of a deplorable but business-as-usual practice concludes the survey of global warming's contemporary repercussions. What comes in the next and final chapter is a different survey: of some of the expected, deleterious consequences of continuing to burn fossil fuels and steps that might alleviate them.

3

WHAT'S NEXT

The preceding chapter reviewed some of the consequences of anthropogenic global warming at the time of writing. This one explores what some of the consequences might be as far ahead as 2100, and what actions have been proposed to prevent, mitigate, or otherwise deal with them.[1]

The Fifth IPCC Report, the 2015 Paris Agreement, and CoP 26

To explore and help understand what the consequences could be, climatologists have used extrapolations from global climate models and contemporary knowledge to create a range of likely GW futures. Such scenarios are presented in the Assessment Reports of the Intergovernmental Panel on Climate Change (IPCC). Beginning with its first one in 1990, these Reports have reviewed extant scientific, technical, and socio-economic information and conclusions concerning the likelihood of anthropogenic climate change, its potential impacts, and options for adaptation and mitigation. The degree of likelihood has increased with each Report, with the fifth one stating unequivocally that global warming is occurring and that it is extremely likely that its dominant cause since

F. S. Levin, *Global Warming: Truth and Consequences*, https://doi.org/10.1007/978-3-031-27023-9_3

1970 is anthropogenic, while the sixth one states that its occurrence is unequivocally anthropogenic.

Unlike weather forecasts, which predict the high and low temperatures on a given day, the Reports replace such forecasts with "projections," each of which delineates the likely ranges of future ΔTs, sea level rises, etc. These ranges are the outcomes of global climate model (GCM) calculations that use different assumptions about CO_2 emissions, economic growth, population growth, and other factors. As stated in Chapter 2, the past comparisons between data and calculations are in sufficient agreement for GCMs to be used for this purpose, which is fortunate, as there are no other means of creating scenarios about future repercussions. The projections in all the Reports but the first are based on global climate model results obtained since the preceding one, and because the GCMs created by different climatologists do not necessarily yield identical numbers, the Reports present a range of results for each scenario.

The fifth IPCC Report (AR5), published in 2013 and 2014, played a significant role in the discussions and conclusions of the highly touted Paris Agreement of December 2015, which was signed by representatives of 195 countries. Its stated goal was limiting ΔT to not more than 2 °C above the value of preindustrial levels by 2100, with a concerted effort to reduce that value to 1.5 °C, which was a demand of vulnerable, developing countries.

That the Agreement was highly touted is due in part to every one of the representatives having acknowledged that anthropogenic global warming (AGW) is occurring, which was a first. To counter the likely consequences of AGW, individual emission reduction goals were made part of the Agreement. Furthermore, the wealthier countries pledged to raise $100 billion a year by 2020 to help poor countries transform their economies. Overall, the Agreement is legally binding, but some elements, including the pledges by individual countries to curb emissions and the promise of financing by some, are not binding. Furthermore, it has no enforcement provisions that could be used to coerce countries to meet their pledges: they are voluntary ones.

The United Nations Framework Convention on Climate Change (UNFCCC) adopted the Agreement in 2015 and asked the IPCC to prepare a report, to be published in 2018, on steps to be taken to implement the goals of the Agreement. Published as the "IPCC Special Report on Global Warming of 1.5 °C," it was approved by the UNFCCC in October 2018. Among its findings were that achieving the ΔT limit of 1.5 °C in 2100 would require a reduction of anthropogenic CO_2 emissions of roughly 45% from the 2010 value by 2030, and to reach net zero by 2050. Net zero means that any anthropogenic emissions still occurring would be balanced by removing the same amount from the atmosphere. This might be done by reforestation or direct capture from the atmosphere, actions that are examined later.

To understand what must be undertaken to achieve this, note that in 2010, 30.6 gigatonnes (Gt) of CO_2 were emitted globally, which was a 5.7% increase over 2009. The 45% reduction to the 2010 value therefore means that in 2030 the emissions should be approximately 16.8 Gt. Given that CO_2 emissions in 2019 were 33.7 Gt and had recovered in 2021 to 33 Gt, this will be an enormously difficult undertaking. In addition, fossil fuels are not the only sources of greenhouse gas emissions (mostly CO_2): they also arise from cement production and land use. An estimate of the global total in 2018 was approximately 44 Gt (with a slight increase in 2019, a small decrease in 2020, and nearly the 2019 level in 2021), so applying the 45% reduction to the 2018 value means a total decrease of greenhouse gases to nearly 25 Gt by 2030. Achieving this is considered later in this chapter, although I note here that as of late 2021, most of the countries that have produced the majority of the emissions have not met their pledged goals, nor has the $100 billion per year been forthcoming. While none of the industrialized countries are yet on track to meet their pledged goals, some others have been.

Since the pledges had not been met, emissions had continued to rise until the occurrence of the COVID-19 pandemic (when they fell but later rebounded), and no major decisions were agreed to at the 2018 and 2019 meetings on further steps to be undertaken, then it seems that US President Donald Trump's decision to withdraw from the Paris Agreement will not have important practical consequences, nor will US President Joe Biden's decision to re-enter it. Why not? Because the emission pledges in the Paris Agreement that are analyzed in the IPCC's 2018 Special Report indicate that these pledges are *not sufficient* to keep ΔT from exceeding 2 °C by 2100: to do so would require more drastic actions than agreed to in Paris.

An opportunity for the world's countries to commit to such drastic actions occurred at the United Nation's 26th Climate Change Conference of the Parties, referred to as CoP 26, held in Glasgow, Scotland in late fall 2021. Delayed one year due to Covid-19, it was attended by an astonishing 30,000 persons, including representatives from 197 countries as well as people from non-governmental organizations (NGOs), trade unions, businesses, lobbyists, and other interested parties, including young people who had hoped for significant progress but came away voicing disappointment yet again.

Nevertheless, some important commitments were made, but they were not as stringent as initially agreed to. I shall review only a few features of the final agreement here; interested readers can find details on the internet.

First, there were continued commitments to cut carbon emissions by nearly half by 2030 and to try to keep ΔT not greater than 1.5 °C by 2100. However, the plans that countries submitted beforehand to achieve this failed to do so, with analyses showing that the end of century warming from them would be 2.4 °C, in line with the finding noted above from the 2015 Paris Agreement. Second, the initially agreed-on program to achieve the foregoing commitments that included the "phasing out" of coal-fired-generating plants was objected to by China and India at essentially the last minute, who demanded that "phase out" be replaced by "phased down" or they would not sign the agreement. This replacement was made. However, no dates were specified for the phase-down, nor were any steps agreed to for doing this. Third, and this is important, even though the agreement included "phase down," there was a strong sense among almost all the countries' representatives that coal needed to be and would be phased out. Finally, and equally important, it was agreed to meet again in 2022 rather than in five years, where further attempts would be made to achieve the commitments noted above. Stay tuned.

In view of this situation and the fact that the pledges made in 2015 were found in the 2018 Special Report to be insufficient to keep ΔT below 2 °C, an obvious question arises: Are there significant differences between the impacts of a ΔT of 1.5 °C and a ΔT of 2 °C? Given that the current increase to 1.1 °C has produced the negative repercussions described in the previous chapter, it should be no surprise that the answer is Yes: the consequences associated with a ΔT of 2 °C are more severe. As noted in the Special Report, these include increases in the number of hot days, of drought, of rainfall deficits, and of heavy precipitation, i.e., the kinds of repercussions discussed in Chapter 2. In addition to these qualitative assessments, there are quantitative ones concerning temperature extremes. The extreme hot days at mid-latitudes are projected to be hotter by 3 °C versus 4 °C, while extreme cold nights are projected to heat up by 4.5 °C versus 6 °C(!) at high latitudes. In addition, the Special Report asserts that with a $\Delta T = 1.5$ °C, 70–90% of coral reefs would be lost, whereas that number goes up to greater than 99% for $\Delta T = 2$ °C by 2100.[2]

And there's more: if $\Delta T = 2$ °C by 2100, extreme heat is almost three times as likely to affect the world's population compared to the lower limit 1.5 °C. There will be more people worldwide subject to severe drought caused by water scarcity for the higher ΔT at 2100: the likely values are approximately 137 million versus 194 million. Species losing more than 50% of their normal range at 2100 are at least twice as likely at the higher limit compared to the lower one; these species include vertebrates. It should thus be clear that the two different values of ΔT do lead to significant differences, ones that partly underlie the (very difficult to attain) goal of ΔT not exceeding 1.5 °C sought by the vulnerable developing countries.[3]

The preceding results come from global climate model-based projections. Described in the next section is the most recent source of such projections available at the time of writing. Its projections for the temperature ranges in the last decades of this century, the amounts of CO_2 emissions in 2100, and some estimates of sea level rise are described below.

The Sixth IPCC Report

That source is the sixth IPCC Assessment Report (AR6), which as noted above, states for the first time that GW is unequivocally anthropogenic. The global climate models used to create its sets of scenarios are known as CMIP6 (where CMIP stands for Coupled Model Intercomparison Projects). As in the past Reports, the different GCMs that were used in AR6 produced different anomalies. Because they are anomalies, it is necessary to identify the base period relative to which they are specified. The ΔTs of Figure 1.3 are relative to the ΔT mean of the period 1880–1920, which is approximately 0 °C. The maximum value of ΔT in Figure 1.3 is approximately 1.3 °C, which occurred in 2020. The website "Updating the Climate Science," administered by Makiko Sato with James Hansen, which is the source for Figure 1.3, also has a graph of the ΔTs relative to the 1951–1980 base period. In it, the maximum value of ΔT is just over 1 °C, which means that the mean value of the 1951–1980 base line is about 0.3 °C higher than that of Figure 1.3. Turning finally to the ΔT ranges of the AR6 scenarios, which are for global surface temperatures, its baseline for is the period 1850–1900, whose ΔT mean value is approximately 0 °C, just like that of Figure 1.3.

The Working Group 1 portion of AR6 (the Physical Science Basis) was published online in August 2021; it includes a Summary for Policy Makers, which is the source of the material described below. The AR6 scenarios are labeled Shared Socioeconomic Pathways (SSPs). Their forcings differ from those of AR5 in how the emissions occur over time, in the mix of CO_2 and other greenhouse gases, and in other factors such as the treatment of aerosols.[4] More importantly, the GCMs that are used to make the relevant projections are much more sophisticated. This should be no surprise given that the climatologists who created them had more than six years to improve their codes—and there are many more of them who have contributed.

The AR6 Summary for Policy Makers includes values for ΔT over several time periods, but only those for the end of century range 2081–2100 are shown in Table 3.1, where the radiative forcings (in W/m^2) are listed in the SSP column. The best estimate values are bracketed by lower and upper limit values; they occur because each of the different groups that contributed to the CMIP6 used their own GCM, which produced results that differed at least somewhat from the others. Despite this, the ranges of the SSP ΔTs are identified as *very likely* ones. Such an identification means a greater degree of confidence in the results than in previous Reports, since the degree of likelihood in this case lies in the interval 90–100%. The major factor underlying this higher degree of confidence is that the AR6 GCMs take account of the various inputs and simulations of the earth's climate system with improved accuracy.

Each of the five sets of ΔT ranges (lower, best, upper) is linked to a single value of the total CO_2 emissions in 2100, with the graph of each of these five single values starting from the total CO_2 emissions of 40 Gt in 2015. For the best-case scenarios of SSP 1.9 and 2.6, the total emissions in 2100 are negative, with the former at roughly -15 Gt and the latter at approximately -10 Gt. Obviously, in these two scenarios procedures to both cut

Table 3.1   The SSP Temperature Anomalies

| SSP | ΔT range (2081–2100)[a] | | |
	Lower Limit	Best Estimate	Upper Limit
1.9	1.0	1.4	1.8
2.6	1.3	1.8	2.4
4.5	2.1	2.7	3.5
7.0	2.8	3.6	4.6
8.5	3.3	4.4	5.7

[a]ΔT values in °C

emissions and remove CO_2 from the atmosphere (discussed later) have succeeded. Nevertheless, that does not happen quickly enough to keep all six of these ΔTs below the limit of 1.5 °C demanded by developing countries in 2015, with the SSP 2.6 upper limit value exceeding the limit of 2 °C that was agreed on in Paris in 2015.

How is it possible that some of these ΔTs can exceed 1.5 °C given the strong negative emissions in 2100? The answer is that enough CO_2 remains in the atmosphere to continue raising global temperatures. This is especially true for the intermediate case of SSP 4.5, for which emissions rise slightly until roughly 2050 but then fall to about 10 Gt in 2100: all three of its ΔT values exceed 2 °C, as do the best estimates of the final three: each is well above the desired 2 °C upper limit of the Paris Agreement. Clearly these are outcomes to be avoided, if at all possible.

SSP 7.0 and 8.5 are the "worst-case" scenarios, ones where emissions reach what can only be regarded as disastrous levels. SSP 7.0 is referred to as the *high emissions* scenario, in which emissions increase almost linearly until doubling to 80 Gt in 2100, whereas in SSP 8.5, the *very high emissions* case, they peak at around 130 Gt in approximately 2080, remain there for about ten years and then fall off to roughly 125 Gt in 2100. Three of the six ΔT values in these worst cases well exceed 4 °C, itself a catastrophic result, one that is described below.

Before turning to that description, the other feature from the Summary I consider is the set of projected global sea level rises. Except for a particular outlier for SSP 8.5 which I'll specify shortly, for each SSP there is a narrow range of outcomes, but since all five lie within a small interval, I'll list only the lower and upper values. Relative to the value for 1900, they are 0.5 meter for SSP 1.9 and 1 meter for SSP 8.5. While a 1-meter rise would be a calamity for many parts of the world, the Carbon

Brief article listed in chapter note 4 mentions that some climatologists have claimed that such a rise is a conservative underestimate, given the likelihood range of 66–100%. For them, 2 meters is a more realistic outcome. Nevertheless, even a 3-foot (not quite a 1 meter) rise is ominous, a point discussed later in this chapter.

Finally, there is the SSP 8.5 outlier noted above. It is a low-likelihood, high impact storyline, including ice-sheet instability processes that results in a rise of 1.75 meters. Not quite as disastrous as 2 meters, but still a major disaster, as so much of the earth's land masses would be under water.[5]

What if $\Delta T = 4\ °C$

While global climate models can produce a range of projected outcomes for temperature, sea level rise, heat waves, etc., it is obvious that they cannot be used to determine which, if any, of the outcomes will occur by the end of the century. Even so, one thing seems certain: if efforts to drastically rein in CO_2 emissions are not implemented, then there can be increases in ΔT, sea level rise, etc., that will have ominous consequences, as indicated in the previous section. To address this, I survey next what some of the repercussions will be for $\Delta T = 4\ °C$ by 2100, the main source for which is an article on it in the Guardian Weekly of May 31, 2019, supplemented by additional information from the internet.

A ΔT of 4 °C would be an extreme result, but Table 3.1 shows that it is no longer an outlier, though its likelihood is not as great as smaller values. Furthermore, $\Delta T = 4\ °C$ is the central value in the World Meteorological Organization's late 2018 projection of a 3 – 5 °C increase by 2100 if the then current trend continued, which it did through the first part of 2020, then decreased, and rebounded in 2021. Note also that the 3 °C lower limit arises from the assumption that the 2015 pledges would materialize—which they had not as of the time of writing. Analogously, as mentioned above, a 2-meter sea level rise has been proposed for 2100; it is also a projected consequence of a $\Delta T = 4$ °C increase. However, and it is a very important however, I am in no way claiming that a ΔT of 4 °C by 2100 will occur. It simply serves as an example of a "what-if," and in that sense is a warning because the projected repercussions are frightening.

The first thing to note is that the negative consequences discussed in the preceding chapter will become much worse in a $\Delta T = 4$ °C at 2100. These involve the decimation of coral reefs; hurricane activity for which the higher ocean temperatures will lead to greater areal extent and amounts of water vapor; more intense monsoons and torrential rainfall in general; the vanishing of Arctic summer sea ice and its overall diminishment; continued ice sheet loss, which in one scenario could increase sea level by over a foot and possibly much higher; and expected extinction of at least one third of animal and plant species.

In addition to the foregoing, the heat and heat waves will be far more severe. While there will be higher if not much higher temperatures encountered in different parts of the globe than those specified in Chapter 2, an augury of how devastating forest fires would be in the $\Delta T = 4$ °C world of 2100 are the incredibly destructive ones that ravaged the west coast region of the USA, Australia, and Siberia noted above. How much worse such fires will be globally is terrifying—assuming that is, that large forests, including the Amazon's, will not have been greatly reduced in size in 2100 by logging and previous fires.

Temperature increases will not only be unequally distributed globally in 2100 for $\Delta T = 4$ °C, in some regions they might be as high as 10 °C. Poorer nations will suffer more than wealthier ones, an analog of which already exists in the USA: black and brown and other minority neighborhoods have been impacted by increased heat more than others. The unequally distributed heat waves will produce a broad equatorial belt with very high humidity and accompanying heat stress, making it uninhabitable for most if not all the year, while north and south of it some of the land will become deserts, even as far north as southern and central Europe. Not only will much of humanity end up living in higher latitudes, which will be the less heat-impacted regions, but the former agricultural areas will be wiped out. The number of hunger refugees will soar far beyond what already exists, which is in the tens of millions.

At least two-thirds of the Asian glaciers whose ice-melts (along with snow and rain) help feed that region's rivers will be lost,[6] leading to widespread water scarcity, food shortages, and energy deficiencies. Upward of two billion people may be affected. Glacier loss in other parts of the world will have similar consequences. Even now water scarcity is a problem, since roughly a quarter of the world's population, including parts of the USA, lives in water-stressed areas, due in part to ground water aquifers being depleted, more quickly in some countries than others. It is thought that hundreds of millions of people will be impacted in Africa alone. In the $\Delta T = 4$ °C world, sea levels are projected to be 20% higher in the tropics, making low-lying islands and coastal cities there uninhabitable and some non-existent. Assuming current population trends continue unabated, that could mean over one billion sea-rise refugees by 2100, with nearly double that as a real possibility.[7]

Given this outline of devastating consequences, it is important to re-emphasize that as appalling as they are, they arise from a scenario whose likelihood of realization cannot be predicted. I have included them to highlight a situation that humanity must do everything in its power to prevent—if it can. Some of the steps that are needed are discussed below, but it is important to remember that they must be politically realizable and achievable technologically.

Nevertheless, some may argue that these projections should be taken with a "grain of salt." After all, they are artifacts of mathematical models that simulate the earth's climate system, ones that don't take all of the system into account, and perhaps should not be relied upon. Indeed, they do come from such models, and as Yogi Berra famously said, "It's tough to make predictions, especially about the future." The counter to this is that such models have done reasonably well in their shorter-term predictions and as noted earlier, they produced long-term projections whose results are in reasonably good agreement with one another even though they come from different models. But the most important counter is that with $[CO_2]$ values having consistently risen,[8] the amount of heat they trap has also risen, thus raising global temperatures. The rapidity of the $[CO_2]$ increases is the greatest in many thousands of years, with current values equal to or greater than in past times when the sea levels were far greater than those projected by the models. It can only get worse as atmospheric CO_2 levels are expected to continue increasing unless potent steps are taken. Furthermore, CO_2 remains in the atmosphere for at least 100 years, so that if at any time emissions were suddenly to cease, their presence in the atmosphere would *maintain T_{av} at its then value if not increase it.* That is why achieving net zero emissions is necessary.

So, what are the kinds of steps that are needed? They are simple to state, but not to implement, and include reducing current emissions, removing CO_2 from the atmosphere as well as capturing and storing it permanently before it can be emitted, and switching over to renewables, among others. In the following sections, I will survey some of them. My treatment will be fairly-broad-brush.

Reducing Current Emissions

There are two items that are relevant when one considers reducing emissions of greenhouse gases: which global economic sectors are involved (e.g., transportation) and what are the sources globally (e.g., oil) that contribute to energy use/consumption. While the US Environmental Protection Agency (EPA) had been a source for this type of information, it was not under the Trump administration and has not been updated quickly enough under the Biden administration at the time of writing, so one must seek other sources for the more recent information.

For the economic sectors, I have used the World Resources Institute report for the year 2016, whose results I have combined into four categories: transportation (15.9%), electricity, heat, and building usages (38.8%), various industrial processes (23.8%), and agriculture, forestry, and land use (18.3%). These may be compared with the numbers for the USA on p. 28, where transportation is noticeably higher and agriculture lower.

For the energy use/consumption, the following results are the most recent available at the website *Updating the Climate Science*, whose data is presented in a pie chart for 2020: oil at 31.2%, coal at 27.2%, natural gas at 24.7%, and renewables including nuclear at 16.9%. These results scarcely differ from those in the 2019 report *Trends in Global CO_2 and Total Greenhouse Emissions* by Olivier and Peters. In the latter report, the percentages of the major greenhouse gases emitted globally were (from 2017 and 2018): CO_2 at 72%, CH_4 at 19%, and N_2O (nitrous oxide) at 6%. China and the USA together emit about 40% of the global total, with China in the unenviable lead. In view of the above set of results, one of the obvious steps to reducing emissions is to close coal-fired energy-generating plants. In many parts of the world this is happening, sometimes by direct government order with the intent to switch to renewables, and in some cases, as in the USA, for bottom line reasons: using natural gas had been much cheaper until late 2021 and early 2022, when its price increased, and coal usage jumped by 17.6%. How that will change in the USA remains to be seen. But many coal plants elsewhere continue to operate, with a number under construction in China, which already has more in use than all other countries combined, as recently reported in the *Guardian*. While it has closed many, those still operating in China could continue doing so for decades, so one must wait and see if China will close many more in order to live up to its 2015 Paris Agreement commitment. Ditto for closings in India: recall "phase down" replacing "phase out" from CoP 26 above. There is, to some extent, a hopeful sign: as reported by the *Guardian* and the *New York Times*, energy generation by solar, wind, and other renewables finally surpassed that produced by coal in the USA in 2019.

Another step toward reducing emissions is for electricity-generating plants to switch to renewables, but so far, for the vast majority (approximately 90%) it is not a priority. The *Guardian*, which reported this, also disclosed the depressing action of the world's biggest investment banks, which have invested over $2.66 trillion into fossil fuel extraction since the 2015 Paris Agreement, even though some have publicly identified the extreme dangers of global warming. Not exactly putting one's money where one's mouth is. More on renewables below.

On the transportation side, switching from gasoline powered vehicles to electric powered ones would certainly help with reductions, since they emit so little CO_2. They are also far more fuel efficient than gasoline powered vehicles whose efficiency varies from about 12% to 30% of the energy obtained from the fuel. The numbers of electric cars and buses were around five million in 2020; one projection is that nearly a third of all vehicles will be electric by 2030, with another estimate claiming that half of all vehicles will be electric by mid-century, with many auto manufacturers undertaking electric vehicle production. There is, unfortunately, a counter to this possibly rosy picture: the craze for sport utility vehicles (SUVs), which were 40% of all vehicles sold globally in 2019, according to a report in the *Guardian*. That same report notes that taken as a unit, they are the seventh biggest emitter of CO_2 globally and are so popular that the US car maker General Motors was planning at one time to have them as 54% of its sales by 2025. Other makers may well follow suit. One can hope that the craze for SUVs will diminish substantially, or that as gasoline-powered vehicles they will become much more fuel efficient, or that they will quickly become electric powered (the fuel efficiency of SUV's is currently about 15% less than gasoline-powered cars, at least in the USA). Where the electricity would come from for powering such vehicles as well as being used commercially is addressed later.

Important as these reducers are, there is another that has been considered to be a possible game-changer: making the cost of emitting CO_2 and other greenhouse gases so expensive that consumers will rebel against using the sources that produce them and switch to others that are either much less or even non-CO_2 emitting, assuming, of course that these "others" exist in sufficient quantities, e.g., all-electric vehicles whose electricity comes from renewables. The way to accomplish this is by governments putting a price on CO_2 emissions to help reach the 2050 goal—at least in developed and developing countries.

One method would be to levy taxes on consumers and any entities whose products generate CO_2. Requiring consumers to pay the tax would seem to be unfair since they are largely passive emitters (excluding owners of passive houses and buildings or electric vehicles), though governments have not shied away from introducing taxes on their inhabitants for other purposes. The alternative is taxing the fuel industries (coal, oil, and gas companies) for their eventual carbon emissions (a "carbon" tax), which would almost certainly increase the prices consumers would pay (this would be an indirect tax on consumers without naming it as such). If the amounts levied by governments were to increase over time, which would lead to higher prices paid by consumers, then it is assumed that their purchasing of the more expensive products would decrease, being replaced by alternatives (electric vehicles, for example). If this went on long enough, some industries might have a difficult time remaining in business (as in the case of US coal companies, but for different reasons) unless they had invested in relevant research and development or in renewable sources of energy—which many organizations, including fossil fuel companies are already pursuing, or claiming to. The desired end-result would be less if not much less CO_2 being emitted. Of course, consumers' needs for some of the associated products would still exist, so other means would be needed to produce them in ways not involving CO_2 emissions. More on this feature at the end of the chapter.

There are essentially two types of carbon taxes: "Emissions Trading Systems (ETS)" also known as "Cap and Trade," and what one might call "ordinary" taxes. A subset of the latter is known as "Fee and Dividend," which is the version I'll explore shortly.

In the former, the amount of emitted CO_2 is initially capped at a chosen maximum. The maximum would be set by national governments, presumably based on the best science known. If a national government did not introduce a cap, then individual governmental entities such as states in the USA could do so. Once a cap has been set, a total number of permits, one for each ton of CO_2 emitted, would be determined. The number distributed either freely or by auction, to each entity whose products produce emissions, would be based on its history: as many permits as needed to meet their historical emissions. A fuel imported from abroad and used to generate CO_2 would count in the historical emissions.

Entities that reduce the resulting emissions more quickly than others would have some permits left over and could trade (i.e., sell) them to those who reduce less quickly. Market forces of one kind or another would determine the cost of the traded permits. Permits could be "retired," thus decreasing the number of permits and driving up the cost of trading permits, making it more expensive to emit. To achieve the overall reduction, for instance 45% by 2030, the cap would be decreased over time. A producer that exceeded the goal set by the number of permits it initially obtained and was unable to buy more would pay a penalty to the governmental agency that set the cap. This penalty would effectively be a tax, but not on consumers. However, since the steps needed to reduce individual emissions would likely be costly, the consumer would end up paying more for whatever goods the producer sells—for example, heating oil or gasoline. A drawback to this scheme concerns entities whose products generate very large amounts of CO_2, who could decide not to reduce the amount as quickly as the permits required and instead pay the penalty from their enormous amount of capital. The effect of this would be to delay achieving the overall reduction of emitted CO_2 by the specified date, allowing more CO_2 to go into the atmosphere, thereby increasing not reducing global warming. Cap and Trade has been implemented in various places with differing degrees of success. Readers can find discussions of this on the internet.

In the Fee and Dividend scheme, a government would levy an initial tax (fee) on a unit of emissions, say per ton. Each year the tax would increase by a given amount, making it more expensive to emit. The tax increase would inevitably be passed on to consumers by increasing the price of whatever product is involved, e.g., heating oil or gasoline, so it seems that the consumer would be stuck with increasingly higher costs. However, the *dividend* portion is intended to offset the higher costs, so that the consumer would not be subject to the increases (or very much of it).

Here's how it is supposed to work. Assume the initial fee is $50 per ton, increasing each year by $20 or $30. Instead of the government keeping the money raised by the fees, it would distribute the money equally to consumers as a dividend, so each consumer (up to a family of two adults and two children, for instance) would be able to offset at least some, if not most or all of the price increases. A drawback to this method of taxation would be setting up and maintaining one or more centers to distribute the dividends, which would likely require a new database of the names and addresses of the consumers. The expense of doing this would offset some of the dividend unless the government subsidized this expense, which seems unlikely, or worse, a government could decide not to distribute the dividend at all.

A problem with this type of tax proposal is an expected reluctance of governments to impose them specifically as a means of reducing CO_2 emissions, especially in the current political climate of the USA and geopolitically elsewhere. How developing countries would benefit is not clear, nor if and how they might be required to participate. In other words, I am not optimistic about achieving this. Nonetheless, it should be attempted.[9]

The foregoing concerns reducing CO_2 emissions. Nevertheless, attention must be paid to those of methane (CH_4), which hit an all-time high in 2021 according to a January 2022 report from the US National Oceanic and Atmospheric Administration (NOAA). These emissions have increased by about 9% since 2000. The main sources are fossil fuel burning and the belching of cows. Switching to electric vehicles and changing the diets of cows would very likely go a long way to alleviate this problem. It is important to do so because CH_4, whose atmospheric lifetime is several decades, is a much more potent absorber of heat than CO_2, whose own atmospheric lifetime is much longer. The technical measure of this potency is its "global warming potential" (GWP), where that of CO_2 is taken as 1 for a 100-year lifetime. The GWP of CH_4 is estimated to be 86 over 20 years (i.e., 86 times more potent than CO_2 for its 100-year lifetime), but it drops to about 34 when a 100-year atmospheric lifetime is analyzed, with each number including what are known as feedback effects that make it more powerful.

But reducing existing CH_4 emissions is not the only concern: another, not immediate but potentially much more harmful were it to occur, is the release of methane from melting permafrost on land and from methane hydrates in areas of Arctic waters 50–100 meters deep. At least 90% of permafrost consists of methane along with rocks, soil, and microbes, but only traces of CO_2, whereas hydrates are essentially frozen methane embedded in a crystal structure of water.

Permafrost is defined as below-ground soil whose temperature has been lower than freezing for at least two years. In the northern hemisphere, its depth can be a meter or less, or as much as 1.5 km (e.g., in Siberia). The northern hemisphere area which permafrost underlies was estimated at one time to be about 25 million square kilometers (25% of the surface), but a 2021 analysis reduced that to about 14 million square kilometers or about 15% of the northern surface, still a very respectable amount. If the frozen ground above it thaws sufficiently, the permafrost can begin to melt, which will lead to two consequences. The lesser one is subsidence and reconfiguration of the surface leading in part to collapse of roads or houses. This has already occurred in parts of Alaska, Siberia, and the Canadian arctic where permafrost is thawing 70 years earlier than had been predicted. The major one is the interaction of microorganisms with the near surface permafrost that frees the CH_4, which can then be emitted: once in the air it can also combine with oxygen forming CO_2. These releases have happened but not yet on a large scale. The concern is thus for the future, when further increases in ΔT as well as arctic fires will lead to significantly greater amounts of methane emissions, leading via positive feedbacks to further increases in ΔT, etc. Emissions from methane hydrates have also been observed, but not in large amounts so far, though that potential exists, as explained in Peter Wadham's book *A Farewell to Ice*.

While the above schemes for reducing emissions are important, another is to capture them before they can enter the atmosphere, for example, from electricity-generating plants or when extracting the fossil fuel from a well. With capturing, an immediate question arises: what is to be done with the extracted gases? The answer is that they must be stored somewhere, permanently, or almost so. This is the problem of sequestration, which is considered in the next section.

Further Steps to Be Taken

One of these further steps concerns efficiency. Not only do vehicles need to be more efficient, buildings should be also. What one might call the gold standard here is the "passive" house or commercial building, which require at most small amounts of heating and cooling and thus have a small greenhouse gas footprint. A relatively recent innovation, they are mostly found in European countries. Many if not most "normal" houses and buildings tend to be fuel-inefficient: they would consume less electricity and/or heat if they were retrofitted, e.g., with either better or new insulation, better windows, white roofs.

But the essential step involves removing CO_2 from the atmosphere, which will be needed to maintain net zero emissions. A seemingly simple way to help accomplish this is by planting a huge number of trees. One proposal calls for a trillion such plantings: not to have lots more leaves to absorb CO_2 that would be released when the trees lose them, but to absorb the carbon via photosynthesis as part of a tree's growth to maturity. It is a proposal that has garnered much attention and comment. However, it is not quite so simple an action, as there are various issues that need to be considered. Among them are which trees to plant and where, possible interference with other land use and its existing entities, availability of water, viability, time to grow to sufficient maturity, and how much carbon could be absorbed. Such planting could help but is far from a panacea—other actions would be more useful if not much more so. Readers interested in this topic might try searching the internet for "pros and cons of planting trees to address global warming"; two of the links I have found useful are Yale Climate Connections and Skeptical Science. In addition, I recommend David Kramer's article "Negative Carbon Dioxide Emissions," published in the January 2020 issue of *Physics Today*, to which I will refer below. But until anything like tree planting might be done, what must be considered is the fact that deforestation is continuing, with one estimate being over 38,000 km^2 of forest lost yearly.

Planting trees to absorb carbon is not the only biological procedure that has been proposed to remove it from the atmosphere. Another "natural" way of absorbing CO_2 is known as "bioenergy with carbon capture and storage" (BECCS). In it, "biomass" is grown, absorbing the carbon by photosynthesis, after which the biomass is harvested and treated by one of several processes that releases the CO_2, which is to be captured and stored. Ideally, the growing will occur without impacting the environment or agriculture, which the US Department of Energy in 2016 claimed could be done safely in the USA. At least 1 Gt of material from agriculture, forestry, waste, and algal substances would be available to produce a large amount of biofuel or biopower, as reported in Kramer's 2020 *Physics Today* article. The process is not as efficient or cheap as using natural gas and has been widely demonstrated in only a few places. It is a method waiting to be further developed and deployed. Information on it is available on the internet.

A more consequential procedure is direct air capture (DAC). Unlike tree planting or BECCS, this is a commercial process involving machinery of some kind. Consequently, determining or at least estimating the costs per amount of CO_2 captured[10] is an essential ingredient of it. However, capturing the gas is just one part of DAC;

how it is to be sequestered to prevent it escaping back to the atmosphere is another, which is considered below.

One type of machinery used to capture the gas is a free-standing tower that contains material which can absorb it chemically; others are more complex, e.g., involving fans to force the air into the relevant receptacle for chemical absorbing or adsorbing. In his most recent Physics Today article on climate change (January 2022), David Kramer examines some of the issues and identifies a few companies involved in DAC, with one of them, Climeworks, a Swiss company operating in Iceland, being identified as the first commercial plant to achieve DAC. An important plus is that it uses geothermal rather than fossil fuels for energy, but, as Kramer notes, it has the capacity to capture only 4000 tonnes of CO_2 per year, injecting it deep underground. There are other companies that have gone into this process, as discussed in the Wikipedia article on it, which presents arguments pro and con. A drawback is that large amounts of energy are needed for the machinery to operate, so that for DAC to be in widespread use, as is necessary to meet the goal of net zero, this energy requirement (at least as is currently envisaged) is a problem that must be solved. DAC has not yet been scaled up globally.

A different procedure that requires sequestration is capturing the gas before it gets into the atmosphere, for example, at electricity- or heat-generating plants, oil wells, ethanol producers, etc. Known either as Carbon Capture and Storage (CCS) or Carbon Capture, Utilization and Storage (CCUS), it is discussed in Kramer's latest article, which identifies several attempts in the USA, including some that have not produced the desired results. Many more details are available in the Wikipedia article on CCS, including capture and storage projects in different countries along with achievements. Like DAC, this is a commercial process, so costs are an aspect here, too. Readers interested in details are recommended to the Wikipedia articles on it and on DAC.

All this finally brings us to the question of sequestration—i.e., of storage of the captured gas. There are several storage procedures that have been used or proposed, among which are injecting the CO_2 directly or in a compressed form deep into the earth's sedimentary rocks; uniting it with building materials such as cement or concrete; and chemically transforming it into minerals of one kind or other. The first of these, geologic storage, is thought to be safe, i.e., with no re-emergence. Kramer's 2020 article notes that only five geologic storage operations existed globally in 2019; the oldest being the North Sea Sleipner gas field, which since 1996 has injected one million tons of CO_2 annually with no leakage. The Global CCS Institute report for 2021 states that there were 27 operating facilities worldwide, with 108 in the project pipeline. The total geologic storage capacity is estimated to be between 20 and 30 trillion tons of CO_2, so there is plenty of storage space. How quickly facilities—and how many— for actual DAC and storage will be developed remains to be seen.[11]

The mineralization procedure is a form of "rock weathering," a natural phenomenon whereby the CO_2 is chemically transformed into different minerals by interacting with rocks that have sufficient amounts of calcium, magnesium, or iron, all in the form of cations, which are elements that have lost one or more electrons and become positively charged. For the mineralization procedure to work, the rocks need to be finely ground, with the amount of CO_2 absorbed chemically depending on the rock. In principle, there is a gigantic amount of mantle rock within a few km of the earth's surface that could be used for storing hundreds of trillions of tons of CO_2. The rocks would have to be mined and then finely ground, with the CO_2 needing to be transported to where the rocks would be placed. This is costly, with estimates ranging from as low as \$20 to hundreds of dollars per ton captured. In addition, a US National Academy of Sciences report on this notes that the procedure would lead to huge volumes of waste that could contaminate water, air, or both, which would need to be dealt with. It remains to be seen, yet again, how this will develop.

Renewables

One of the essential activities needed to achieve ΔT not becoming too great by 2100 is to replace fossil fuel sources of energy by what are known as renewables, ones that do not emit greenhouse gases and whose energy is intrinsically renewed. They include hydropower, wind farms on land and/or in the ocean, solar panel farms on land, bioenergy, and geothermal.[12] For me (and for at least some climatologists) nuclear power must be part of this mix—even though many environmentalists revile it. It is sometimes called a non-renewable, though as a non-CO_2 emitter, it would fit into the renewable category. I will consider it last.

One crucial question is how much of the world's energy consumption comes from renewables and nuclear. As of the time of writing the latest definite results are for the year 2020: renewables at 12.6% and nuclear at 4.3%, which are the values from the website *Updating the Climate Science*. In detail, renewables breakdown into hydropower at 6.88%, wind energy at 2.54%, solar power at 1.37%, biofuels at 0.68%, and geothermal and biomass sources at 1.12%.[13] This is a large increase in renewables from the preceding year, as stated in *The BP Statistical Review of World Energy* for 2019, where renewables are at 5% while nuclear is 0.1% higher than the 2020 value.[14]

In addition to these definite results, there is a prediction from the International Energy Agency (IEA) in its report of December 1, 2021, that when finally assessed, renewable power in 2021 will set a record, and that by 2026 renewable electricity capacity will grow by 60% over the amount in 2020 while renewables overall will account for 95% of the increase in global power capacity. This is a positive upside; the downside is the concomitant prediction that such growth is only 50% of what is needed to achieve net zero by 2050; to reach it growth needs to be 80% faster. Yet in an era when banks are continuing to fund fossil companies, coal is still an energy source in several countries, and China is committed to reach net zero only in 2060, it is far from clear that achieving net zero by 2050 will be possible. Stay tuned over the next several decades.

In addition to the downside noted above, a drawback to solar and wind is that their sources are intermittent: solar farms absorb nothing at night and wind farms require wind. It is therefore important to be able to store the electricity they produce when there is no solar radiation to absorb and/or no wind is available. An obvious choice to accomplish this is batteries. While ones with sufficiently large capacities are actively being investigated, none seem available for this purpose at the time of writing. Since the growth of both solar and wind electricity generation has been increasing and will continue to do so, the development of much larger battery carrying power will be needed even more.

For land-based wind farms in the USA, there is another problem: most of the states where the wind blows the strongest (e.g., the Midwest) are not nearly as populated as much of the rest of the country, so how will its electricity be transmitted to more populous regions? Answer: via the National Grid, which is how all electricity, so far, has been transmitted to consumers, with individual power lines leading off the Grid as needed to send the electricity to different communities.

The Grid consists of 120,000 miles (193,000 km) of lines operated by roughly 500 individual companies, according to a recent Wikipedia article. They operate independently within each of the three regions of the contiguous USA, designated Western, Eastern, and Texas, and they exchange little power, though it is possible to do so across the regions.

There are issues concerning the Grid. For one, how easily will wind-powered electricity be transmitted from one region to another? It is also inefficient, though not nearly as much so as in some other countries, and it is aging, as the catastrophic 2021 failure of the Grid in Texas has demonstrated, though it was strengthened enough in 2022 that the failure involved far fewer outages. Nevertheless, it is not clear that the Grid as it exists now can easily handle the intermittent electric flows from wind (or solar), or how the associated storage batteries (when available) will fit into its infrastructure. With the expected increase in electric vehicles, will the current incarnation of the Grid be able to supply the extra electricity needed for them, either directly into houses or into what is likely to be a large set of charging stations around the country? Because of its segmented organization in the USA, such problems will be dealt with on a piecemeal basis, ones that could be treated more easily if it were truly a "national" Grid. However, transforming it into a national entity seems highly unlikely. Discussions of the Grid and its problems can be found on the internet, for example, in Wikipedia, in the June 20, 2020, article on Vox, and in links that are reached by typing "problems with the US National Grid" into a search engine.

From here I turn to a consideration of nuclear power. Many people look it on very negatively, in part because of the fears that accidents—natural or human-made—will release radioactive materials, including radiation itself. That aspect of the horrific bombing of Hiroshima and Nagasaki in World War II has not been forgotten in the USA and elsewhere (especially, of course, in Japan), even though it is so-called ancient history.

In recent times, as far as is known, there have been only two accidents that released a substantial amount of radioactive material: Chernobyl and Fukushima, whereas the Three Mile Island one in the USA emitted a miniscule amount (despite the fears it aroused, especially with the release soon after of the film "The China Syndrome"). The former accidents occurred because of very poor construction of the Chernobyl facility that led to a calamity over a very large area, while the sea wall that was supposed to protect Fukushima was far too low for the supposed once in a hundred-year surge that overtopped it. Coupled with these events is the real concern that nuclear power plants built on or next to fault lines could experience disasters when an earthquake occurs. And for all plants built next to rivers or lakes, necessary for cooling water purposes, either an earthquake or a hurricane could also cause a disaster of one kind or other (a meltdown of the reactor core, for example).

Finally, there is the problem of waste disposal from the standard type of reactor in use around the world, and the associated risk of nuclear bomb material being accessed from that waste. Very careful waste disposal is required because some of the radioactive elements have lifetimes of many thousands of years. For France, roughly 75% of whose electricity is nuclear generated, waste disposal has not been a problem because it is recycled. These are all serious reasons that have given rise to fear and loathing concerning nuclear power. Is there, however, another side to this story? The answer in my opinion is Yes.

I'll start with the present. Nuclear power has been and is being generated by what are known as light water reactors (LWRs),[15] which have been used in various countries for a long while, with some new ones having been built or under construction in various nations; Russia has supposedly been exporting some. According to the World Nuclear Association's website, as of February 2022 about 10% of the world's electricity was generated by nuclear power plants, of which there were about 440 along with approximately 220 research reactors, which produce isotopes for medical and industrial uses. In the USA, since storage of radioactive waste from nuclear plants had been proposed under Yucca Mountain in Nevada but denied, all such wastes have been placed in sealed containers and distributed among 80 sites in 35 states. Unlike France, the USA does not recycle spent nuclear fuel, so the amount stored will continue to grow at a rate of about 2000 tons per year. No leaks or thefts have been reported (though it seems to me unlikely that if any had occurred that it would be announced publicly).

So much for the present. Why nuclear power should be looked on more favorably is its possible future incarnations, since a LWR is not the only type of reactor. A very different type is the molten salt reactor, a noteworthy example being the liquid thorium fluoride reactor (LFTR),[16] which is *safe against a meltdown, produces far less plutonium, and whose radioactive wastes decay in far shorter lifetimes than its LWR sibling*, among other features. Despite claims that a few countries are developing them, they are not in widespread use at present, though that may well change in the near to medium future. The cost of constructing one is estimated to be considerably less than for a LWR, so this, along with the features just noted, are reasons that LFTRs ought to be a nuclear power source of the future. Unfortunately, being new and essentially unutilized seems to mitigate against their adoption, at least in the USA, as its nuclear industry appears very reluctant to innovate, despite the estimated lower costs. In addition to LFTRs, several other types have been proposed; none seems yet to be operating as power sources. Readers interested in these topics can find much information on the internet. For an overview of LFTRs, pros and cons, the Wikipedia article is a good start. I especially recommend accessing the Hargreaves and Moir article in the American Scientist that is listed in Footnote 8 of the Wikipedia article: it is a clear and concise discussion of LFTRs.

In light-water and molten-salt reactors energy is produced via the process called *nuclear fission*. An example of this process is a neutron (n) colliding with and being absorbed by the nucleus (N) of a very heavy atom, after which the new nucleus (consisting of $N + n$) either releases one or more neutrons or radioactively fissions into two or more lighter nuclei; the energy produced powers electricity-generating plants as in the case of LWRs.

In contrast to the above two fission reactors, a third process that in the relatively near future could be used as a nuclear power source involves *nuclear fusion*. Fusion is the opposite of fission, in that two very light nuclei collide and then combine, producing energy and a new, slightly heavier, stable nucleus (and possibly a lighter one). The simplest example of fusion is a neutron combining with a proton (p) to produce a new nucleus called a deuteron (consisting of $n + p$) along with energy in the form of a gamma ray.

Fusion reactions occur in the central core of stars, where the energy needed to begin them is supplied by the core's extremely high temperature, around 15 million °C for stars like the sun. These reactions yield a large amount of energy, so if a fusion reactor could become a practical reality, it would, among other things, be an energy source that would avoid the radioactive by-products of fission reactors. That has been a decades-long goal. What has prevented it from occurring is the giant amount of energy needed to start the fusion reactions, and a container that would confine the reacting nuclei to its interior, allowing the process to continue generating more energy than is needed to initiate it.

There have been many attempts to achieve this, carried out in different countries, sometimes collaboratively, but success has essentially remained elusive until some recent developments, ones that led me to write above "a third process that in the relatively near future could be used as a nuclear power source involves *nuclear fusion*." They appear so promising that some government entities and/or venture capitalists have invested in them. The reason they appear promising is the claim by several companies to have available by 2025 either a demonstration one or a commercialized one. If so, this would be a major advance in the ability to supply energy in a non-greenhouse gas emitting way and therefore could be a game changer in the long run. Interested readers can learn more by typing "startup fusion companies" into a search engine or going to the article "fusion power" in Wikipedia.[17] Of course, it remains to be seen how this develops, but it is worth being aware of.

Geoengineering

Up to this point, my discussion has focused on reducing greenhouse emissions, on removing CO_2 from the atmosphere and sequestering it, and on the use of renewables. There is, however, a radically different approach to the global warming problem, which I'll briefly consider next: reducing solar insolation via geoengineering so that less solar energy reaches the earth, which in turn means less re-radiated IR to be trapped in the atmosphere and thus less global warming.[18]

A variety of methods for blocking the insolation have been proposed; a common drawback is unintended consequences. One method is to inject sunlight-reflecting aerosols (sulfates or nitrates) into clouds or the stratosphere. Large amounts would be needed, and since their lifetime in the lower atmosphere is short, this procedure would require constant repetition. While the cost would presumably be large and agreement from countries where the injection would occur would be needed, a bigger problem is that aerosols reaching the earth from clouds or the stratosphere would lead to aerosol-contaminated air, the breathing of which is toxic to humans.

An analogous procedure is marine cloud brightening (MCB), which involves continuously spraying sea-water droplets into the thin, low-level clouds that cover about a fourth of the globe's oceanic surface. These droplets would help the clouds reflect the insolation, presumably in sufficient amounts to achieve the same goal as aerosol seeding. The procedure is described in Peter Wadham's book (listed in the Bibliography), which cites the kind of investigations needed to determine unintended consequences. As reported in the April 17, 2020, issue of The Guardian, it had a trial of sorts in March 2020, one that was designed to test whether the droplet delivery system would work to help protect portions of Australia's Great Barrier Reef from further bleaching, and not to test whether the method itself is effective. The droplet delivery system did work. While this procedure is of continuing interest, it remains for the future to determine if a hugely scaled up version will be effective as a solar radiation reflector, and at what cost and drawbacks.

The only other geoengineering procedure I'll consider is what Burton Richter (see the Bibliography) refers to as a sunshade (i.e., a mirror) in space that would reflect incoming sunlight. At present it remains a theoretical proposal with no implementation planned. Costs are estimated to be very large—hundreds of billions of dollars—and since it would need to reflect continuously, it would need to be in orbit. That feature could negatively impact some of the very many satellites already stationary or orbiting the earth. A good discussion of this proposal is in the Wikipedia article on it. Other geoengineering proposals can be found by searching the internet.

COVID-19

While the long-term effect on greenhouse gas emissions from the COVID-19 pandemic is unknown, the short-term ones, those for 2020 and 2021, have been identified. As noted previously, it led to a decrease in CO_2 emissions in 2020 that was rebounded from in 2021. Unfortunately, the global decrease in emissions did not lead to a decrease in the yearly average global temperature for 2020: as noted in Chapter 2, the 2020 value of ΔT is as high as in 2016, without a corresponding super El Niño. This result reinforces the need for so much still be done to prevent a large ΔT increase in 2100, especially with CO_2 emissions surging in 2021.

Not only did CO_2 emissions rebound in 2021, but global methane emissions have hit record levels and have continued to rise. The main causes have been fossil fuel burning, an increase in the number of cows (for whom COVID-19 is clearly not a health hazard), and leaks from what are known as methane hot spots in different parts of the earth.

In view of the fact that some countries are relaxing COVID-induced restrictions, including lockdowns, and that automobile and air travel has not only resumed but in some places exceeded previous amounts, it seems to me likely that both emissions and ΔT increases will continue, despite new COVID variants that almost certainly will occur. In other words, COVID will become, to some extent, like the flu: new versions that will need to be dealt with but will not impact GW. Thus, the requirement of continued reduction of emissions throughout the 2020 decade to reach the 45% noted at the beginning of this chapter will remain. The USA's rejoining of the Paris Agreement could be a catalyst for this if two things occur. First, that the Inflation Reduction Act of 2022 is fully implemented and not later repealed in the USA and second, that that action induces other countries, especially China, to undertake drastic reductions in this necessity. Time will tell.

Some Final Comments

In contrast to the preceding remarks, my first comment is both tongue in cheek and somewhat serious. Suppose that oil is finally phased out. There is then the possible problem of finding non-CO_2-emitting sources of the many consumer goods whose production involves oil in some way. Since there is a vast amount (thousands) of these, with many involving plastics, I have selected a small, arbitrary number from among them so the reader will have some notion of what they are: ink, bicycle tires, denture adhesives, food preservatives, wheels, toothpastes, eyeglasses, telephones, clothing, antiseptics, credit cards, aspirin, toilet seats, heart valves, ballpoint pens, vitamin capsules, and lipstick. This list includes only a few plastics, since they are already a major pollutant that needs to be phased out. The need for innovation here is immense; it seems to me that it will be a formidable task to achieve. Of course, could providing the oil needed for these products be enough to keep some oil companies in business without burning any oil, just extracting and sending it to be appropriately processed? If not, the above problem will remain to be solved.

My second comment is that there are some hopeful signs. These include the increasing and widespread use of wind and solar, whose costs have declined considerably, the development of much lower cost batteries for electric vehicles, which could be charged directly from home if solar panels are attached to the house, and the real possibility of fusion power. Announcements by some governments including the EU itself indicate a desire to strongly reduce emissions by 2030, while the IEA projects that 95% of all new energy will be generated by "clean" sources by 2025. President Xi has announced that China will achieve net zero emissions by 2060. In early 2021 in the USA, the automobile maker General Motors announced that it would produce only zero-emission vehicles by 2035 (which other car manufacturers have cited as partial goals); the CEO of the American Petroleum Institute acknowledged that the threat of climate change is real and existential and that the API supports efforts by industry and the federal government to address this issue; the Biden administration working to reverse the Trump administration's roll back of climate change regulations put in place by the Obama administration as well as passing major climate change initiatives in August 2022. Whether they can achieve the claimed effects remains to be seen, especially in view of the likelihood that if Republicans control the government, they will roll back these initiatives. Nevertheless, all these efforts do not go far enough to achieve the reductions needed to rein in the warming, as outlined in above.

In other words, much more needs to be done, and it needs to be done as much as possible globally and not just in a few developed counties. The question is how much can be achieved, e.g., will the claims made at CoP 26 be implemented and if so, when?

My last comment is that as an educator I have tried to present what I think are the main issues that need to be dealt with—at least for developed countries. There remain those countries that are not in this class, some of whom are poor. Their typical sources of heat are carbon-emitting fuels, e.g., wood or coal. How will their contributions to global warming be dealt with? Over time this could be done by continuing to install renewables like wind and solar power. While in some places this is being done rapidly, it seems likely that it will not be done uniformly unless the large sums of money that were promised at the December 2015 meeting are forthcoming. As of the time of writing, the amounts that have been given are not sufficiently close to what was promised, though some countries have done better than others. Putting aside the humanitarian reason to improve the needed standard of living for inhabitants of these poorer counties (which should not be put aside, though many persons in developed countries believe that such help is someone else's problem), the funds need to be provided to help save humankind from the least deleterious effects that are likely to happen in 2100 if at least a $\Delta T = 2\ °C$ increase occurs.

So, where are we and what should be done, in view of the situation described in this last section of this chapter and in portions of Chapter 2? My answer to the first part of the question is that we seem to be at a stalemate because of the inertia and/or opposition in many of the current social, political, and economic environments. If this could be overcome, then some of the steps that have been outlined must be taken to try to limit ΔT to $2\ °C$.

Therefore, the real question is: Can the necessary actions be taken and quickly enough? I leave you with my considered answer: it is unlikely in view of the stated intention in the USA and elsewhere to continue burning fossil fuels and even drilling for more as a response to the Russian invasion of Ukraine. Even more chilling in this regard is a report in the Guardian Weekly of 20 May 2022 on the immense amount of money that, according to its investigators, fossil fuel companies and governments are planning to invest in new drilling and fracking through 2030. Because of these intentions and plans, I am concerned not only for the planet's inhabitants in general but also for the selfish reason that my grandchildren and their children as well as those of yours will very likely be living in a ravaged global environment, one that will hardly resemble the world we inhabit now. Avoiding this ought to be a sufficient motivation to take needed action, especially in view of the dire warnings from the April 2022 IPCC Report. Despite my pessimism, I am not so despairing as to entertain the tiniest of hopes that some of the needed actions can occur, at least over time. Yet I will not make any bets on it, not only because of the Guardian Weekly report of May 20, 2022, but also by the report in the Guardian of July 21, 2022, that for the last 50 years the gas and oil industry has made a *daily* profit of $2.8 billion!! In view of such sums, it is hard not to wonder if the above acknowledgement from the API CEO is a form of greenwashing. As with much I have noted, only time will tell.

NOTES

Introduction

1. For refutation and debunking of rejecters' spurious claims, see the web sites Skeptical Science and RealClimate.
2. The phrase "climate change" is often used as a replacement for if not an equivalent to "global warming." However, global warming is the "parent" whose "children" are the disastrous climate changes described in this book, and I am therefore using "global warming" throughout the book.

Chapter 1

1. Three years prior to Tyndall's investigations, the American scientist Eunice Foote carried out experiments on the effect of the sun's rays on different gases. Among her conclusions was that adding CO_2 to the atmosphere could increase the earth's temperature. Her paper on this was published in 1856 in the American Journal of Science and Arts, and though it was reported in newspapers in the US and Europe, it remained unrecognized by the scientific community until it was unearthed by the geologist Ray Sorenson in 2010, who publicized it, thereby helping to gain the recognition Foote should long have had. For further information see the Wikipedia article about her.

© The Editor(s) (if applicable) and The Author(s), under exclusive license to Springer Nature Switzerland AG 2023
F. Levin, *Global Warming: Truth and Consequences*,
https://doi.org/10.1007/978-3-031-27023-9

2. These numbers are available at various web sites, for example Wikipedia.

3. Monthly ppm values can be obtained by using an internet search engine and at various web sites such as the US National Oceanic and Atmospheric Administration (NOAA) and the Scripps Institution of Oceanography.

4. Figure 1.2 was the latest available at the time of writing this chapter that showed both the Mauna Loa and Antarctica results on the same graph. Later Keeling curves can be accessed at the Scripps Institution of Oceanography website and the internet.

5. Relative humidity is the amount of water vapor in the air expressed as a percentage of the total amount it can hold at the same temperature. Specific humidity is approximately equal to the ratio of the mass of water vapor in a volume of air to the mass of dry air in the same volume. As the air temperature rises, it increases, as noted above

6. Svante Arrhenius' crude mathematical simulation of the earth's climate system is the forerunner of the modern ones, the global climate or global circulation models. They involve at least a million lines of computer code and have been part of the rejecters' attacks on the AGW paradigm, as discussed in the next chapter.

7. Isotopes have the same number of protons and electrons but different numbers of neutrons in their nuclei. The source for this analysis is the RealClimate post of 18 December 2004.

8. Among the many sources I have relied on for details are the books by Oreskes and Conway, and by Mann, whose chapter notes have been very helpful for documenting various claims. Readers interested in a listing

of the rejecters' claims and brief rebuttals to them can find them on the Skeptical Science website (see the Bibliography). The books by Oreskes and Conway and by Mann also contain accounts of some of the attempts (many successful) to spread uncertainty, for example by ExxonMobil. Additional details and references in this regard are listed in a Wikipedia article as well as in many websites available from internet searches.

9. For further refutation and debunking of rejecters' spurious claims see the web site RealClimate.

10. Since T_{av} is about 14 °C, it is advantageous to plot the anomalies ΔT, as they are small enough that they would not stand out so prominently on a plot of T_{av} versus time. Note that I am just describing such graphs here; actual ones are displayed later in the chapter.

11. As noted previously, a global climate model (GCM), sometimes referred to as global circulation model, is a simulation of the earth's climate system whose features (atmospheric components, etc.) are described by as much relevant physics the computer(s) carrying out the calculations can handle. GCMs are discussed later in the chapter.

12. See the internet for relevant information including the graphs. Although this is somewhat ancient history and is an action that Michaels much later downplayed, it dramatically exemplifies one of the many rejecter tactics.

13. The medieval warm period occurred in the interval 1000–1200 in the North Atlantic land regions. It was warm enough in England that wine was produced. The graph that appeared in the 1990 IPCC repot was drawn by hand.

14. The IPCC (the Intergovernmental Panel on Climate Change) was created in 1988 by the World Meteorological Organization (WMO) and the United Nations Environment Programme (UNEP). Its purpose is to assess "the scientific, technical and socioeconomic information relevant for the understanding of the risk of human-induced climate change. It bases its assessment mainly on published and peer reviewed scientific technical literature." Details of some of its reports are discussed in Chapter 3.

15. The Climate Research Unit (CRU) is another source of yearly average global temperature data. To access it search on "crudata.uea.ac.uk/cru/data/temperature/". The data that appears in the graphs in this chapter are from the web site "Updating the Climate Science," as indicated earlier.

16. The revised data is included in the next section's graph. These data problems led to a proposal that heat entering the oceans should be used to identify GW, since that heat had been measured and shown to be steadily increasing.

17. This video can be accessed on You Tube's "This is Not Cool," by videographer Peter Sinclair, who originated it. It is also available at "climatecrocks.com" by clicking on their Overview, scrolling down on "Archives" to their January 2016 post and then scrolling down to "Retrieval Algorithm," the fourth video on the January 29 page.

18. Ice ages are caused by decreases in the amount of sunlight reaching the earth that are due to changes in both the eccentricity and the precession of the earth's elliptical orbit, and to the tilt of its axis of rotation. These changes form the Milankovic cycle; it can last approximately 100 thousand years, with much shorter interglacials, for example the one we are living in now having begun around 12 thousand years ago. See Wikipedia for details about the cycle.

19. Figure 1.3 is the graph of temperature anomalies as of January 2022. Like its many monthly predecessors, it is displayed at the data-rich web site *Updating the Climate Science*, which Makiko Sato maintains with James Hansen. Starting in 2021, the anomaly graphs were reconfigured so that the linear trend is specified by a solid green line running from 1970 to 2015, while the portion after 2015 is represented by a dotted green line. The change was suggested by Hansen to reflect the expectation that global warming has been accelerating; the dotted portion will indicate that by being above the extension of the solid line. At present it is not, but that could change, even going below it. Note, by the way, that 1970 is usually considered to be the year in which average global temperature increases can unequivocally be identified as anthropogenic. See *Updating the Climate Science* for later versions of the ΔT anomaly graphs.

20. I will not be discussing the Southern Oscillation portion of ENSO, which refers to the oscillation in air pressure between eastern and western portions of the Pacific Ocean, since its behavior is not crucial for describing the direct temperature roles played by El Niño and La Niña. These phenomena, by the way, alter rainfall patterns globally, leading to drought in some places and heavy inundations in others.

21. Forcings are entities that influence the earth's energy balance and thus its temperature. They are measured in W/m^2; the insolation of this chapter is an example. In the present case the forcings are greenhouse gases, black carbon (for example soot), ozone, solar irradiance, snow albedo, stratospheric water vapor and aerosols, land use, indirect aerosol effects, and reflective tropospheric aerosols. Radiative forcing (RF) is the difference between the solar radiation absorbed by the earth and that which is radiated back into space.

22. See the RealClimate post of 4 December 2019.

Chapter 2

1. Most of the numbers and commentary in this chapter are from various internet web sites (e.g., CNN) as well as articles in the New York Times, The Guardian, The New Yorker, and other periodicals, so unless a source is specified, readers are encouraged to use their search engine to seek out verification on the internet if they need it.

2. See the web site Updating the Climate Science for graphs of sea level rise.

3. Eventually the Newport Naval Base, the largest one in the USA, will no longer be viable, which raises the inevitable question of where it—and many other military bases—might be located. A further point in this general regard is that the flooding due to sea level rise can be, and in some cases has been, exacerbated by the sinking of land at and near the coastline. Such subsidence has occurred, for example, along the US east coast and California, and more dramatically in Louisiana, which has lost over 5000 km² (over 2000 miles²) since roughly 1930, a loss that is continuing. Most of it is in Terrebonne and Plaquemines Parishes, and while efforts are being undertaken to stem it and replenish some of the land, the long-range forecast is a pessimistic one, with predictions of New Orleans becoming an isolated island. And that is if, in the longer run, it survives, since not only has the Gulf of Mexico come much closer to it, allowing for greater hurricane devastation, but portions of the city, itself at or well below sea level, are sinking at a rate of about one-half a foot per decade, according to satellite data.

4. See comments in Jeff Goodell's book "The Water Will Come," listed in the Bibliography.

5. A good source is the July 2020 special issue of the National Geographic and accompanying map.

6. There is a context to this that may not be common knowledge: the World Health Organization concluded in 2017 that 2.1 billion people did not have access to clean drinking water, while 4.5 billion, more than half of the earth's human population, lacked adequately protected water for sanitation. The former situation improved to some extent in 2020, in that the number decreased to 2 billion, while the number lacking adequately protected water for sanitation improved to 3.54 billion, still a devastatingly large number, about 46% of the world's population.

7. Graphs of Greenland and Antarctic ice sheet losses are available at the web site Updating the Climate Science.

8. This number is an indirect measure of the decrease in depth of the ice. Since the change was about 43% for the period 1976 to 1999, and global warming has continued, the estimate of a 73% decrease for the longer period is consistent with the 43%. The larger decrease is a combination of the reduction of Arctic ice area and the fact that global warming has changed Arctic ice from thicker multi-year ice to significantly thinner first-year ice. The ice loss over many years is dramatically displayed by searching on "Piomas Arctic Death Spiral" and NASA's website.

9. My main source for the information on leaks and earthquakes resulting from fracking is Rachel Maddow's book *Blowout*, listed in the Bibliography; see also earthquaketrack.com/p/united-states/oklahoma/recent and USGS articles on Oklahoma earthquakes.

Chapter 3

1. What is likely to occur and when, and how it may be dealt with, may be influenced by the COVID-19 pandemic. It had caused a significant reduction in emissions due to major economic downturns, which emissions later rebounded. How it may affect global warming in the future is discussed at the end of the chapter.

2. These consequences are labeled by degrees of confidence, almost all of which are medium or high. The confidence level associated with the temperature values is high.

3. In addition to the Special Report on Global Warming of 1.5 °C, the IPCC issued two other Special Reports after it: "On Climate Change and Land," and "On the Ocean and Cryosphere in a Changing Climate." Interested readers may consult them for their insights into these topics.

4. For more details concerning Shared Socioeconomic Pathways than presented here, readers can enter "CMIP6: the next generation of climate models explained," into their search engine. It will take them to an article with this title on the website Carbon Brief written by the climatologist Zeke Hausfather and updated before the appearance of the Working Group 1 portion of the sixth IPCC report. I have not quoted any of the actual numbers in this article because, as Hausfather points out, they were produced by only a portion of the CMIP6 GCMs and gave results that do not agree with those from the full CMIP6 set found in the Summary for Policy Makers.

5. What I have presented from the Summary for Policy Makers omits important comments that its authors include concerning not only the various ΔTs, but also

many aspects of the Summary. Interested readers are encouraged to read the Summary for these fascinating and cautious remarks concerning the results in it. Very informative comments about it can be found in the 13 August 2021 post at RealClimate. As noted in the text, the information from the Summary for Policy Makers presented on pp. 79 – 82 is from the Working Group 1 portion of AR6 that was published in August 2021. As I was creating the Index for this book (which was much later revised), the Working Group 2 portion of AR6 was published online in February 2022. Its main topics are impacts on ecosystems, biodiversity, and human communities, and how adaptations to climate change can be dealt with. As there is overlap between its impacts and those of Chapter 2 as well as portions of this chapter, plus much more detail (just as in the Summary from Working Group 1), I am leaving it to interested readers to access it. As with this book, be prepared for disturbing results.

6. Measurements in 2019 have shown that since 1970 the Tibetan plateau had already lost 25% of its ice.

7. Readers interested in consequences of different sea level rises can find discussions and maps on the internet. For example, typing "Effect of an n-foot sea rise in different parts of the world" into a search engine brings up many links, where n is the reader's choice: for instance, $n = 3$, or for truly horrific results (but not likely in 2100) $n = 10$.

8. One might think that this should not be a concern due to COVID-19, which led to a decrease in emission levels in 2020 of about 7%. Despite this, as noted in the preceding chapter, the December 2020 value of CO_2 was 415 ppm, not a significant decrease, while the April 2021 value was 421 with January 2022 at

417, while in August 2022 it was 420, as noted earlier. See further comments later in the chapter.

9. Discussions of these schemes, including pros and cons, can be found on the internet, for example Wikipedia. One treatment I found useful is David Kramer's article "Should carbon emissions be taxed or capped or traded," published in the December 2019 issue of *Physics Today*, available online to non-members during COVID-19.

10. Such details are immaterial to the purposes of this book, so interested readers should consult the internet and/or Kramer's article, if it remains available online.

11. As reported in the New York Times in March 2021, one proposal in this direction is being undertaken in Teesside, England, supported by the oil giant BP. It involves a new plant powered by natural gas whose emissions would be captured, fed into pipes, and then carried well out into the North Sea. There they would be pumped into porous rocks, where, presumably, they would be buried permanently. Other major oil companies have signed on. It is hoped to use this pipeline burial procedure in other places in the world, where success would mean that fossil fuels could continue to be a source for energy generation, so it is no surprise that oil giants are behind it. An obvious question is whether this sequestering will be permanent and if not, for how long. Time, again, will tell.

12. I am not including what has been called "clean-burning coal," as it does not yet exist, despite claims to the contrary.

13. Because solar is now increasing globally more than twice as fast as wind, it is estimated that solar consumption of energy will surpass that of wind during the 2020 decade, while both together will likely overtake hydro in five years (there are not many places on earth to construct more dams).

14. The year 2019 was a first for two interesting developments: renewables surpassed nuclear in electricity generation globally, while in the USA the percentage from all the renewables but hydro surpassed the percentage of energy supplied by coal: coal had fallen by 15% to its lowest level since 2003, whereas renewables had increased by 1%. It was the first time in 130 years that coal fell behind renewables in the US. In contrast, in 2019 both oil and gas energy consumption continued to increase globally, even though oil's share was down by 0.2% compared to 2018. This continuing increase occurred when fossil fuel use was supposed to be declining, as per the 2015 Paris Agreement.

15. It is almost amusing that the amount of radioactivity emitted from LWR nuclear plants is about a factor of 600,000 *less* than the amount occurring naturally on the planet and is approximately 73,500 *less* than what occurs in the typical human body, according to Burton Richter's book (listed in the Bibliography).

16. Information on molten salt reactors is available on the internet by typing that name into your search engine. Two sources dealing with its possible future developments are the website of the *World Nuclear Association* and the *energy.gov* portion at the *US Department of Energy* website. As a historical note, LFTRs were shelved in the USA after LWRs had been chosen to power nuclear submarines.

17. The two companies that have made the 2025 claims are *General Fusion* for the demonstration and *Commonwealth Fusion Systems* for commercialization. Information on each is available on their websites.

18. DAC plus sequestration is also an example of geoengineering, as is the proposal to seed the ocean with iron, which is intended to increase the number of phytoplankton, allowing for more CO_2 to be ingested by

them. Eventually they would sink to the ocean floor, both removing this additional CO_2 from the atmosphere and storing it, presumably indefinitely except for sudden upwellings. Experiments testing this method have not had significant success at the time of writing.

BIBLIOGRAPHY

Archer, David. 2009. *The Long Thaw*. Princeton, NJ: Princeton University Press.

Berners-Lee, Mike. 2021. *There Is No Planet B*. Cambridge University Press.

Bowen, Mark. 2008. *Censoring Science*. New York, NY: Hyperion.

Broecker, Wallace, and Robert Kunzig. 2008. *Fixing Climate*. New York, NY: Hill and Wang.

Climate Central. 2012. *Global Weirdness*. New York, NY: Pantheon Books.

Goodell, Jeff. 2017. *The Water Will Come*. New York, NY: Little, Brown and Company.

Gore, Al. 2006. *An Inconvenient Truth*. Emmaus, PA: Rodale Books.

Hayhoe, Katherine. 2021. *Saving Us*. New York, NY: One Signal Publishers/Atria.

Hansen, James. 2009. *Storms of My Grandchildren*. New York, NY: Bloomsbury USA.

Hazen, Robert M. 2012. *The Story of Earth*. New York, NY: Penguin Group.

Kolbert, Elizabeth. 2006. *Field Notes from a Catastrophe*. New York, NY: Bloomsbury.

Kolbert, Elizabeth. 2014. *The Sixth Extinction*. New York, NY: Henry Holt and Company.

Kolbert, Elizabeth. 2021. *Under a White Sky*. New York, NY: Crown.

Linden, Eugene. 2006. *The Winds of Change*. New York, NY: Simon and Schuster.

Lynas, Mark. 2008. *Six Degrees*. Washington, DC: National Geographic Society.

Maddow, Rachel. 2019. *Blowout*. New York, NY: Crown.

© The Editor(s) (if applicable) and The Author(s), under exclusive license to Springer Nature Switzerland AG 2023
F. Levin, *Global Warming: Truth and Consequences*,
https://doi.org/10.1007/978-3-031-27023-9

Mann, Michael E. 2012. *The Hockey Stick and the Climate Wars.* New York, NY: Columbia, University Press.

Mann, Michael E. 2021. *The New Climate War.* New York, NY: Public Affairs.

Mann, Michael E., and Tom Toles. 2016. *The Madhouse Effect.* New York, NY: Columbia, University Press.

Martin, Richard. 2012. *Super Fuel.* New York, NY: Palgrave MacMillan.

Monbiot, George. 2007. *Heat.* Cambridge, MA: South End Press.

Oreskes, Naomi, and Erik Conway. 2014. *The Collapse of Western Civilization.* New York, NY: Columbia University Press.

Oreskes, Naomi, and Erik Conway. 2019. *Merchants of Doubt.* New York, NY: Bloomsbury, Press.

Otto, Shawn. 2016. *The War on Science.* Minneapolis, MN: Milkweed Editions.

Pooley, Eric. 2010. *The Climate War.* New York, NY: Hyperion.

Richter, Burton. 2014. *Beyond Smoke and Mirrors*, 2nd ed. New York, NY: Cambridge, University Press.

Roberts, Callum. 2012. *The Ocean of Life.* New York, NY: Penguin Group.

Schmidt, Gavin, and Joshua Wolfe. 2009. *Climate Change.* New York, NY: W. W. Norton and Company.

Seidel, Amy. 2009. *Early Spring.* Boston, MA: Beacon Press.

Speth, James Gustave. 2004. *Red Sky at Morning.* New Haven, CT: Yale University Press.

Vallis, Geoffrey K. 2012. *Climate and the Oceans.* Princeton, NJ: Princeton University Press.

Wadhams, Peter. 2017. *A Farewell to Ice.* Oxford, UK: Oxford University Press.

Wallace-Wells, David. 2020. *The Uninhabitable Earth.* New York, NY: Tim Duggan Books.

Weart, Spencer R. 2008. *The Discovery of Global Warming.* Cambridge, MA: Harvard University Press.

Newspapers and Periodicals

Physics Today.
The Guardian.
The Guardian Weekly.
The New York Times.
The New Yorker.

Selected Websites (An asterisk indicates a data source)

BP Statistical Review of World Energy*.
Carbon Brief*.
Climate Central*.
Fourth National Climate Assessment.
Global CCS Institute.
Global Climate Change.
Global Wind Energy Council*.
Intergovernmental Panel on Climate Change (IPCC)*.
International Energy Agency (IEA)*.
RealClimate.
Scripps Institution of Oceanography*.
SkepticalScience*.
Union of Concerned Scientists.
Updating the Climate Science—Columbia University Earth Institute*.
US Environmental Protection Agency (EPA)*.
US National Aeronautical and Space Administration (NASA)*.
US National Oceanic and Atmospheric administration (NOAA).
Vox.
Wikipedia*.
World Meteorological Organization*.
World Nuclear Association*.
World Resources Institute.
Yale Program on Climate Change Communication.

Figure and Table Sources

Figure 1.2. and the Cover Image Courtesy of Dr. Robert Monroe, Scripps Institution of Oceanography at the University of California San Diego

Figure 1.3. Courtesy of Dr. Makiko Sato, Updating the Climate Science—Columbia University Earth Institute

Figure 1.4. Courtesy of Dr. James Hansen, Dangerous Anthropogenic Interference: 2004 Public Lecture, University of Iowa

Figure 1.5. Courtesy of Dr. James Hansen, *Storms of my Grandchildren*

Table 1.4. Courtesy of Dr. Makiko Sato, Updating the Climate Science—Columbia University Earth Institute

Table 1.5. Courtesy of Dr. Makiko Sato, Updating the Climate Science—Columbia University Earth Institute

Table 1.6. Courtesy of Barton Paul Levenson, how good are climate models now: https://bartonlevenson.com/ModelsReliable.html

Table 3.1. The material following it on CO_2 and sea level rise: from IPCC, 2021: Summary for Policymakers. In: Climate Change 2021: The Physical Science Basis. Contribution of Working Group I to the Sixth Assessment Report of the Intergovernmental Panel on Climate Change [Masson-Delmotte, V., P. Zhai, A. Pirani, S. L. Connors, C. Péan, S. Berger, N. Caud, Y. Chen, L. Goldfarb, M. I. Gomis, M. Huang, K. Leitzell, E. Lonnoy, J. B. R. Matthews, T. K. Maycock, T. Waterfield, O. Yelekçi, R. Yu, and B. Zhou (eds.)]. In Press

© The Editor(s) (if applicable) and The Author(s), under exclusive license to Springer Nature Switzerland AG 2023
F. Levin, *Global Warming: Truth and Consequences*,
https://doi.org/10.1007/978-3-031-27023-9

INDEX

© The Editor(s) (if applicable) and The Author(s), under exclusive
license to Springer Nature Switzerland AG 2023
F. Levin, *Global Warming: Truth and Consequences*,
https://doi.org/10.1007/978-3-031-27023-9